PHYSICS IN MY GENERATION

MAX BORN

PHYSICS IN
MY GENERATION

 Springer-Verlag New York Inc.

PREFACE

THE idea of collecting these essays occurred to me when, in the leisure of retirement, I scanned some of my own books and found that two of the more widely read show a startling change of attitude to some of the fundamental concepts of science. These are *Einstein's Theory of Relativity* of 1921 and the American edition of *The Restless Universe* of 1951. I have taken the introduction of the former as the first item of this collection, the postscript to the latter as its last. These books agree in the relativistic concept of space and time, but differ in many other fundamental notions. In 1921 I believed—and I shared this belief with most of my contemporary physicists—that science produced an objective knowledge of the world, which is governed by deterministic laws. The scientific method seemed to me superior to other, more subjective ways of forming a picture of the world—philosophy, poetry, and religion; and I even thought the unambiguous language of science to be a step towards a better understanding between human beings.

In 1951 I believed in none of these things. The border between object and subject had been blurred, deterministic laws had been replaced by statistical ones, and although physicists understood one another well enough across all national frontiers they had contributed nothing to a better understanding of nations, but had helped in inventing and applying the most horrible weapons of destruction.

I now regard my former belief in the superiority of science over other forms of human thought and behaviour as a self-deception due to youthful enthusiasm over the clarity of scientific thinking as compared with the vagueness of metaphysical systems.

Still, I believe that the rapid change of fundamental concepts and the failure to improve the moral standards of human society are no demonstration of the uselessness of science in the search for truth and for a better life.

The change of ideas was not arbitrary, but was forced on the physicists by their observations. The final criterion of truth is the agreement of a theory with experience, and it is only when all attempts to describe the facts in the frame of accepted ideas fail that new notions are formed, at first cautiously and reluctantly, and then, if they are experimentally confirmed, with increasing confidence. In this way the classical philosophy of science was transformed into the modern one, which culminates in NIELS BOHR's Principle of Complementarity.

To illustrate this process I have selected some of my popular writings covering the period of 30 years which lies between the publication dates of the books mentioned above, and have framed them by the introduction to the first and the postscript to the second. Some of the articles are only loosely connected with the main theme, such as one on the minimum principles in physics, several discussions of EINSTEIN's work, and a modest attempt at autobiography. The remaining articles deal with the philosophical background of physics and its revolutionary changes during my lifetime. There are many repetitions which could not be avoided without spoiling the inner structure of the articles; but I think that each treatment of a problem illuminates it from a different angle, though all of them are given from my personal point of view.

I hope that the collection may transmit to the reader something of the adventurous spirit of a great period of physics.

ACKNOWLEDGMENT

The author, and the publishers, would like to express thanks to those concerned for permission to reproduce material first published elsewhere. In every case details of first publications are given at the head of each article.

CONTENTS

INTRODUCTION TO 'EINSTEIN'S THEORY OF RELATIVITY' (1921)

The world is not presented to the reflective mind as a finished product. The mind has to form its picture from innumerable sensations, experiences, communications, memories, perceptions. Hence there are probably not two thinking people whose picture of the world coincides in every respect.

When an idea in its main lines becomes the common property of large numbers of people, the movements of spirit that are called religious creeds, philosophic schools, and scientific systems arise; they present the aspect of a chaos of opinions, of articles of faith, of convictions, that resist all efforts to disentangle them. It seems a sheer impossibility to find a thread that will guide us along a definite path through these widely ramified doctrines that branch off perchance to recombine at other points.

What place are we to assign to EINSTEIN's theory of relativity, of which this book seeks to give an account? Is it only a special part of physics or astronomy, interesting in itself but of no great importance for the development of the human spirit? Or is it at least a symbol of a particular trend of thought characteristic of our times? Or does it itself, indeed, signify a 'world-view' (Weltanschauung)? We shall be able to answer these questions with confidence only when we have become acquainted with the content of EINSTEIN's doctrine. But we may be allowed to present here a point of view which, even if only roughly, classifies the totality of all world-views and ascribes to EINSTEIN's theory a definite position within a uniform view of the world as a whole. The world is composed of the ego and the non-ego, the inner world and the outer world. The relations of these two poles are the object of every religion, of every philosophy. But the part that each doctrine assigns to the ego in the world is different. The importance of the ego in the world-picture seems to me a measure according to which we may order confessions of faith, philosophic systems, world-views rooted in art or science, like pearls on a string. However enticing it may be to pursue this idea through the history of thought, we must not diverge too far from our theme, and we shall apply it only to that special realm of human thought to which EINSTEIN's theory belongs—to natural science.

Natural science is situated at the end of this series, at the point where the ego, the subject, plays only an insignificant part; every advance in the mouldings of the concepts of physics, astronomy and chemistry denotes a further step towards the goal of excluding the ego. This does not, of course, deal with the act of knowing, which is bound to the subject, but with the finished picture of Nature, the basis of which is the idea that the ordinary world exists independently of and uninfluenced by the process of knowing.

The doors through which Nature imposes her presence on us are the senses. Their properties determine the extent of what is accessible to sensation or to intuitive perception. The further we go back in the history of the sciences, the more we find the natural picture of the world determined by the qualities of sense. Older physics was subdivided into mechanics, acoustics, optics and theory of heat. We see the connections with the organs of sense, the perceptions of motion, impressions of sound, light, and heat. Here the qualities of the subject are still decisive for the formation of concepts. The development of the exact sciences leads along a definite path from this state to a goal which, even if far from being attained, yet lies clearly exposed before us: it is that of creating a picture of nature which, confined within no limits of possible perception or intuition, represents a pure structure of concepts, conceived for the purpose of depicting the sum of all experiences uniformly and without inconsistencies.

Nowadays mechanical force is an abstraction which has only its name in common with the subjective feeling of force. Mechanical mass is no longer an attribute of tangible bodies but is also possessed by empty spaces filled only by ether radiation. The realm of audible tones has become a small province in the world of inaudible vibrations, distinguishable physically from these solely by the accidental property of the human ear which makes it react only to a definite interval of frequency numbers. Modern optics is a special chapter out of the theory of electricity and magnetism, and it treats of the electro-magnetic vibrations of all wave-lengths, passing from the shortest γ-rays of radioactive substances (having a wavelength of one hundred millionth of a millimeter) over the X-(Röntgen) rays, the ultraviolet, visible light, the infra-red, to the longest wireless (Hertzian) waves (which have a wave-length of many kilometers). In the flood of invisible light that is accessible to the mental eye of the physicist, the material eye is almost blind, so small is the interval of vibrations which it converts into sensations. The theory of heat, too, is but a special part of mechanics and electrodynamics. Its fundamental concepts of absolute temperature, of energy, and of entropy belong to the most subtle logical constructions of exact science, and, again, only their name still carries a memory of the subjective impression of heat or cold.

Inaudible tones, invisible light, imperceptible heat, these constitute the world of physics, cold and dead for him who wishes to experience living Nature, to grasp its relationships as a harmony, to marvel at her greatness in reverential awe. GOETHE abhorred this motionless world. His bitter polemic against NEWTON, whom he regarded as the personification of a hostile view of Nature, proves that it was not merely a question of an isolated struggle between two investigators about individual questions of the theory of colour. GOETHE is the representative of a world-view which is situated somewhere near the opposite end of the scale suggested above (constructed according to the relative importance of the ego), that is, the end opposite to that occupied by the world-picture of the exact sciences. The essence of poetry is inspiration, intuition, the visionary comprehension of the world of sense in symbolic forms. The source of poetry is personal experience, whether it be the clearly conscious perception of a sense-stimulus, or the powerfully represented idea of a relationship or connection. What is logically formal and rational plays no part in the world-picture of such a type of gifted or indeed heaven-blessed spirit. The world as the sum of abstractions that are connected only indirectly with experience is a province that is foreign to it. Only what is directly presented to the ego, only what can be felt or at least represented as a possible experience is real to it and has significance for it. Thus to later readers, who survey the development of exact methods during the century after GOETHE's time and who measure the power and significance of GOETHE's works on the history of natural science by their fruits, these works appear as documents of a visionary mind, as the expression of a marvellous sense of one-ness with (Einfühlung) the natural relationships, but his physical assertions will seem to such a reader as misunderstandings and fruitless rebellions against a greater power, whose victory was assured even at that time.

Now in what does this power consist, what is its aim and device?

It both takes and renounces. The exact sciences presume to aim at making objective statements, but they surrender their absolute validity. This formula is to bring out the following contrast.

All direct experiences lead to statements which must be allowed a certain degree of absolute validity. If I see a red flower, if I experience pleasure or pain, I experience events which it is meaningless to doubt. They are indubitably valid, but only for me. They are absolute, but they are subjective. All seekers after human knowledge aim at taking us out of the narrow circle of the ego, out of the still narrower circle of the ego that is bound to a moment of time, and at establishing common ground with other thinking creatures. First a link is established with the ego as it is at another moment, and then with other human beings or gods. All religions, philosophies, and sciences have been evolved for the purpose of expanding the ego to the wider community that 'we' represent. But the ways of doing this are different. We are again confronted by the chaos of contradictory doctrines and opinions. Yet we no longer feel consternation, but order them according to the importance that is given to the subject in the mode of comprehension aimed at. This brings us back to our initial principle, for the completed process of comprehension is *the* world-picture. Here again the opposite poles appear. The minds of one group do not wish to deny or to sacrifice the absolute, and they therefore remain clinging to the ego. They create a world-picture that can be produced by no systematic process, but by the unfathomable action of religious, artistic, or poetic means of expression in other souls. Here faith, pious ardour, love of brotherly communion, but often also fanaticism, intolerance, intellectual suppression hold sway.

The minds of the opposite group sacrifice the absolute. They discover—often with feelings of terror—the fact that inner experiences cannot be communicated. They no longer fight for what cannot be attained, and they resign themselves. But they wish to reach agreement at least in the sphere of the attainable. They therefore seek to discover what is common in their ego and in that of the other egos; and the best that was there found was not the experiences of the soul itself, not sensations, ideas, nor feelings, but abstract concepts of the simplest kind—numbers, logical forms; in short, the means of expression of the exact sciences. Here we are no longer concerned with what is absolute. The height of a cathedral does not, in the special sphere of the scientist, inspire reverence, but is measured in meters and centimeters. The course of life is no longer experienced as the running out of the sands of time, but is counted in years and days. Relative measures take the place of absolute impressions. And we get a world, narrow, one-sided, with sharp edges, bare of all sensual attraction, of all colours and tones. But in one respect it is superior to other world-pictures: the fact that it establishes a bridge from mind to mind cannot be doubted. It *is* possible to agree as to whether iron has a specific gravity greater than wood, whether water freezes more readily than mercury, whether Sirius is a planet or a star. There may be dissensions, it may sometimes seem as if a new doctrine upsets all the old facts, yet he who has not shrunk from the effort of penetrating into the interior of this world will feel that the regions known with certainty are growing, and this feeling relieves the pain which arises from solitude of the spirit, and the bridge to kindred spirits becomes built.

We have endeavoured in this way to express the nature of scientific research, and now we can assign EINSTEIN's theory of relativity to its category.

In the first place, it is a pure product of the striving after the liberation of the ego, after the release from sensation and perception. We spoke of the inaudible tones, of the invisible light, of physics. We find similar conditions in related sciences, in chemistry which asserts the existence of certain (radioactive) substances, of which no one has ever perceived the smallest trace with any sense directly—or in astronomy, to which we refer below. These 'extensions of the world,' as we might call them, essentially concern sense-qualities. But everything takes place in the space and the time which was presented to mechanics by its founder, NEWTON.

Now, EINSTEIN's discovery is that this space and this time are still entirely embedded in the ego, and that the world-picture of natural science becomes more beautiful and grander if these fundamental concepts are also subjected to relativization. Whereas, before, space was closely associated with the subjective, absolute sensation of extension, and time with that of the course of life, they are now purely conceptual schemes, just as far removed from direct perception as entities, as the whole region of wave-lengths of present-day optics is inaccessible to the sensation of light except for a very small interval. But just as in the latter case, the space and time of perception allow themselves to be ordered, without giving rise to difficulties, into the system of physical concepts. Thus an objectivization is attained, which has manifested its power by predicting natural phenomena in a truly wonderful way. We shall have to speak of this in detail in the sequel.

Thus the achievement of EINSTEIN's theory is the relativization and objectivization of the concepts of space and time. At the present day it is the final picture of the world as presented by science.

PHYSICAL ASPECTS
OF
QUANTUM MECHANICS*

[First Published in *Nature,* Vol. 119, pp. 354-357 (1927).]

The purpose of this communication is not to give a report on the present status of quantum mechanics. Such a report has recently been published by W. HEISENBERG, the founder of the new theory (*Die Naturwissenschaften,* 45, 989, 1926). Here we shall make an attempt to understand the physical significance of the quantum theoretical formulae.

At present we have a surprisingly serviceable and adaptable apparatus for the solution of quantum theoretical problems. We must insist here that the different formulations, the matrix theory, DIRAC's non-commutative algebra, SCHRÖDINGER's partial differential equations, are mathematically equivalent to each other, and form, as a whole, a single theory. This theory enables us to compute the stationary states of atoms and the corresponding radiation, if we neglect the reaction of the radiation on the atoms; it would seem that in this respect we have nothing more to wish for, since the result of every example in which the calculations are carried out agrees with experiment.

This question, however, of the possible states of matter does not exhaust the field of physical problems. Perhaps more important still is the question of the course of the phenomena that occurs when equilibrium is disturbed. Classical physics was entirely concerned with this question, as it was almost powerless toward the problem of structure. Conversely, the question of the course of phenomena practically disappeared from the quantum mechanics, because it did not immediately fit into the formal developments of the theory. Here we shall consider some attempts to treat this problem on the new mechanics.

In classical dynamics the knowledge of the state of a closed system (the position and velocity of all its particles) at any instant determines unambiguously the future motion of the system; that is the form that the principle of causality takes in physics. Mathematically, this is expressed by the fact that physical quantities satisfy dif-

---

*Extension of a paper read before Section A (Mathematics and Physics) of the British Association at Oxford on August 10th, 1926. Translated by Mr. ROBERT OPPENHEIMER. The author is very much obliged to Mr. OPPENHEIMER for his careful translation.

ferential equations of a certain type. But besides these causal laws, classical physics always made use of certain statistical considerations. As a matter of fact, the occurrence of probabilities was justified by the fact that the initial state was never exactly known; so long as this was the case, statistical methods might be, more or less provisionally, adopted.

The elementary theory of probability starts with the assumption that one may with reason consider certain cases equally probable, and derives from this the probability of complicated combinations of these. More generally: starting with an assumed distribution (for example, a uniform one, with equally probable cases) a dependent distribution is derived. The case in which the derived distribution is entirely or partly independent of the assumed initial distribution is naturally particularly important.

The physical procedure corresponds to this: we make an assumption about the initial distribution, if possible, one about equally probable cases, and we then try to show that our initial distribution is irrelevant for the final, observable, results. We see both parts of this procedure in statistical mechanics: we divide the phase space into equally probable cells, guided only by certain general theorems (conservation of energy, LIOUVILLE's theorem); at the same time we try to translate the resulting space-distribution into a distribution in *time*. But the ergodic hypothesis, which was to effect this translation, and states that every system if left to itself covers in time its phase space uniformly, is a pure hypothesis and is likely to remain one. It thus seems that the justification of the choice of equally probable cases by dividing the phase space into cells can only be derived *a posteriori* from its success in explaining the observed phenomena.

We have a similar situation in all cases where considerations of probability are used in physics. Let us take as an example an atomic collision—the collision of an electron with an atom. If the kinetic energy of the electron is less than the first excitation potential of the atom the collision is elastic: the electron loses no energy. We can then ask in what direction the electron is deflected by the collision. The classical theory regards each such collision as causally determined. If one knew the exact position and velocity of all the electrons in the atom and of the colliding electron, one could compute the deflection in advance. But unfortunately we again lack this information about the details of the system; we have again to be satisfied with averages. It is usually forgotten that in order to obtain these, we have to make an assumption about equally probable configurations. This we do in the most 'natural' way by expressing the co-ordinates of the electron in its initial path (relative to the nucleus) in terms of angle variables and phases, and by treating equal phase intervals as equally probable. But this is only an assumption, and can only be justified by its results.

The peculiarity of this procedure is that the microscopic co-ordinates are only introduced to keep the individual phenomena at least theoretically determinate. For practical purposes they do not exist: the experimentalist only counts the number of particles deflected through a given angle, without bothering about the details of the path; the essential part of the path, in which the reaction of the atom on the electron occurs, is not open to observation. But from such numerical data we can draw conclusions about the mechanism of the collision. A famous example of this is the work of RUTHERFORD on the dispersion of α-particles; here, however, the microscopic co-ordinates are not electronic phases, but the distance of the nucleus from the original path of the α-particle. From the statistics of the dispersion, RUTHERFORD could prove the validity of COULOMB's law for the reaction between the nucleus and the α-particle. The microscopic co-ordinate had been eliminated from the theoretical formula for the distribution of the particles over different angles of deflection.

We thus have an example of the evaluation of a field of force by counting, by statistical methods, and not by the measurement of an acceleration and NEWTON's second law.

This method is fundamentally like that which makes us suspect that a dice is false if one face keeps turning up much more often than every sixth throw; statistical considerations indicate a torque. Another example of this is the 'barometer formula.' Of course, we can derive this dynamically, if we regard the air as a continuum and require equilibrium between hydrodynamical pressure and gravity; but actually pressure is only defined statistically as the average transport of momentum in the collisions of the molecules, and it is therefore not merely permissible but also fundamentally more sound to regard the barometer formula as a counting of the molecules in a gravitational field, from which the laws of the field may be derived.

These considerations were to lead us to the idea that we could replace the Newtonian definition of force by a statistical one. Just as in classical mechanics we concluded that there was no external force acting if the motion of the particle was rectilinear, so here we should do so if an assembly of particles was uniformly distributed over a range. (The choice of suitable co-ordinates leads to similar problems on both theories.) The magnitude of a force, classically measured by the acceleration of a particle, would here be measured by the inhomogeneity of an assembly of particles.

In the classical theory we are of course faced with the problem of reducing the two definitions of force to one, and that is the object of all attempts at a rational foundation of statistical mechanics; we have tried to make clear, though, that these have not been altogether successful, because in the end the choice of equally probable cases cannot be avoided.

With this preparation we turn our attention to quantum mechanics. It is notable that here, even historically, the concept of *a priori* probability has played a part that could not be thrown back on equally probable cases, for example, in the transition-probabilities for emission. Of course this might be merely a weakness of the theory.

It is more important that formal quantum mechanics obviously provides no means for the determination of the position of particles in space and time. One might object that according to SCHRÖDINGER a particle cannot have any sharply defined position, since it is only a group of waves with vague limits; but I should like to leave aside this notion of 'wave-packets,' which has not been, and probably cannot be, carried through. For SCHRÖDINGER's waves move not in ordinary space but in configuration space, that has as many dimensions as the degrees of freedom of the system ($3N$ for N particles). The quantum theoretical description of the system contains certain declarations about the energy, the momenta, the angular momenta of the system; but it does not answer, or at least only answers in the limiting case of classical mechanics, the question of where a certain particle is at a given time. In this respect the quantum theory is in agreement with the experimentalists, for whom microscopic co-ordinates are also out of reach, and who therefore only count instances and indulge in statistics. This suggests that quantum mechanics similarly only answers properly-put statistical questions, and says nothing about the course of individual phenomena. It would then be a singular fusion of mechanics and statistics.

According to this, we should have to connect with the wave-equations such a picture as this: the waves satisfying this equation do not represent the motion of particles of matter at all; they only determine the possible motions, or rather states, of the matter. Matter can always be visualised as consisting of point masses (electrons, protons), but in many cases the particles are not to be identified as individuals, e.g., when these form an atomic system. Such an atomic system has a discrete set of states; but it also has a continuous range of them, and these have the remarkable property that in them a disturbance is propagated along a path away from the atom, and with finite velocity, just as if a particle were being thrown out. This fact justifies, even demands, the existence of particles, although this cannot, in some cases as we have said, be taken too literally. There are electromagnetic forces between these particles (we neglect for the moment the finite velocity of propagation); they are, so far as we know, given by classical electro-dynamics in terms of the positions of the particles (for example, a Coulomb attraction). But these forces do not, as they did classically, cause accelerations of the particles; they have no direct bearing on the motion of the particles. As intermediary there is the wave field: the forces determine the vibrations of a certain function ϕ that depends on the positions of all the particles (a function in configuration space), and determine them because the coefficients of the differential equation for ψ involve the forces themselves.

A knowledge of ψ enables us to follow the course of a physical process in so far as it is quantum-mechanically determinate: not in a causal sense but in a statistical one. Every process consists of elementary processes, which we are accustomed to call transitions or jumps; the jump itself seems to defy all attempts to visualize it, and only its result can be ascertained. This result is, that after the jump, the system is in a different quantum state. The function ψ determines these transitions in the following way: every state of the system corresponds to a particular characteristic solution, an *Eigenfunktion,* of the differential equation; for example, the normal state the function ψ_1, the next state ψ_2, etc. For simplicity we assume that the system was originally in the normal state; after the occurrence of an elementary process the solution has been transformed into one of the form

$$\psi = c_1\,\psi_1 + c_2\,\psi_2 + c_3\,\psi_3 \ldots,$$

which represents a superposition of a number of *eigenfunktions* with definite amplitudes $c_1, c_2, c_3, \ldots$. Then the squares of the amplitudes, $c_1{}^2, c_2{}^2 \ldots$, give the probability that after the jump the system is in the 1, 2, 3, state. Thus $c_1{}^2$ is the probability that in spite of the perturbation the system remains in the normal state, $c_2{}^2$ the probability that it has jumped to the second, and so on.* These probabilities are thus dynamically determined. But what the system actually does is not determined, at least not by the laws that are at present known. But this is nothing new, for we saw above that the classical theory—for example, for the collision problem— only gave probabilities. The classical theory introduces the microscopic co-ordinates which determine the individual process, only to eliminate them because of ignorance by averaging over their values; whereas the new theory gets the same results without introducing them at all. Of course, it is not forbidden to believe in the existence of these co-ordinates; but they will only be of physical significance when methods have been devised for their experimental observation.

This is not the place to consider the associated philosophical problems; we shall only sketch the point of view which is forced upon us by the whole of physical evidence. We free forces of their classical duty of determining directly the motion of particles and allow them instead to determine the probability of states. Whereas before it was our purpose to make these two definitions of force equivalent, this problem has now no longer, strictly speaking, any sense. The only question is why

---

* We may point out that this theory is *not* equivalent to that of BOHR, KRAMERS, and SLATER. In the latter the conservation of energy and momentum are purely statistical laws; on the quantum theory their *exact* validity follows from the fundamental equations.

the classical definition is so useful for a large class of phenomena. As always in such cases, the answer is: because the classical theory is a limiting case of the new one. Actually, it is usually the 'adiabatic' case with which we have to do: i.e., the limiting case where the external force (or the reaction of the parts of the system on each other) acts very slowly. In this case, to a very high approximation

$$c_1^2 = 1, \ c_2^2 = 0, \ c_3^2 = 0 \ldots,$$

that is, there is no probability for a transition, and the system is in the initial state again after the cessation of the perturbation. Such a slow perturbation is therefore reversible, as it is classically. One can extend this to the case where the final system is really under different conditions from the initial one; i.e., where the state has changed adiabatically, without transition. That is the limiting case with which classical mechanics is concerned.

It is, of course, still an open question whether these conceptions can in all cases be preserved. The problem of collisions was with their help given a quantum mechanical formulation; and the result is qualitatively in full agreement with experiment. We have here a precise interpretation of just those observations which may be regarded as the most immediate proof of the quantized structure of energy, namely, the critical potentials, that were first observed by FRANCK and HERTZ. This abrupt occurrence of excited states with increasing electronic velocity of the colliding electron follows directly out of the theory. The theory, moreover, yields general formulae for the distribution of electrons over the different angles of deflection, that differ in a characteristic way from the results that we should have expected classically. This was first pointed out by W. ELSASSER (*Die Naturwissenschaften*, Vol. 13, p. 711, 1925) before the development of the general theory. He started with DE BROGLIE's idea that the motion of particles is accompanied by waves, the frequency and wave-length of which is determined by the energy and momentum of the particle. ELSASSER computed the wave-length for slow electrons, and found it to be of the order of 10^{-8} cm., which is just the range of atomic diameters. From this he concluded that the collision of an electron with an atom should give rise to a diffraction of the DE BROGLIE waves, rather like that of light which is scattered by small particles. The fluctuation of the intensities in different directions would then represent the irregularities in the distribution of the deflected electrons. Indications of such an effect are given by the experiments of DAVISSON and KUNSMANN (*Phys. Rev.*, Vol. 22, p. 243, 1923), on the reflection of electrons from metallic surfaces. A complete verification of this radical hypothesis is furnished by DYMOND's experiments on the collisions of electrons in helium (*Nature*, June 13, p. 910, 1925).

Unfortunately, the present state of quantum mechanics only allows a qualitative description of these phenomena; for a complete account of them the solution of the problem of the helium atom would be necessary. It therefore seems particularly important to explain the above-mentioned experiments of RUTHERFORD and his co-workers on the dispersion of α-particles; for in this case we have to do with a simple and completely known mechanism, the 'diffraction' of two charged particles by each other. The classical formula which RUTHERFORD derived from a consideration of the hyperbolic orbits of the particles is experimentally verified for a large range; but recently BLACKETT has found departures from this law in the encounters between α-particles and light atoms, and has suggested that these might also be ascribed to diffraction effects of the DE BROGLIE waves. At present only the preliminary question is settled, of whether the classical formula can be derived as a limiting case of quantum mechanics. G. WENTZEL (*Zeit. f. Phys.*, Vol. 40, p. 590, 1926) has shown that this is in fact the case. The author of this communication has, furthermore, carried through the computation for the collision of electrons on the hydrogen atom, and arrived at formulae which represent simultaneously the collisions of particles of arbitrary energy (from slow electrons to fast α-particles). As yet this has only been carried out for the first approximation, and so gives no account of the more detailed diffraction effects. This calculation thus yields a single expression for the Rutherford deflection formula and the cross section of the hydrogen atom for electrons in the range studied explicitly by LENARD. The same method leads to a calculation of the probability of excitation of the H-atom by electronic collision, but the calculations have not yet been completed.

It would be decisive for the theory if it should prove possible to carry the approximation further, and to see whether it furnishes an explanation of the departures from the Rutherford formula.

Even, however, if these conceptions stand the experimental test, it does not mean that they are in any sense final. Even now we can say that they depend too much on the usual notion of space and time. The formal quantum theory is much more flexible, and susceptible of much more general interpretations. It is possible, for example, to mix up co-ordinates and momenta by canonical transformations, and so to arrive at formally quite different systems, with quite different wave functions ψ. But the fundamental idea of waves of probability will probably persist in one form or another.

ON THE MEANING
OF PHYSICAL THEORIES

[A lecture given at the public session of November 10, 1928. Nachrichten der Gesellschaft der Wissenschaften zu Göttingen, Geschäftliche Mitteilungen 1928–29.]

Whoever regards in a detached way the development of the exact sciences must be impressed by two contradictory features. On the one hand, the whole of natural science exhibits a picture of continuous and healthy growth, of unmistakable progress and construction, evident as much in its inward deepening as in its outward application to the technological mastery of Nature. Yet, on the other hand, one observes at not infrequent intervals the occurrence of upheavals in the basic concepts of physics, actual revolutions in the world of ideas, whereby all our earlier knowledge seems to be swept away, and a new epoch of investigation to be inaugurated. The abrupt changes in the theories are in marked contrast to the continuous flow and growth in the realm of well ascertained results. We may give a few examples of such convulsions of theories. Consider the most ancient and most venerable branch of physical science, astronomy, and the ideas concerning the stellar universe, whose course we can follow through thousands of years. At first, the Earth is at rest, a flat disc at the center of the Universe, round which the constellations move in orderly procession. Then, almost simultaneously with the realization of the earth's size and spherical shape, comes the Copernican system of the Universe, placing the sun in the center and allotting to the earth only a subordinate place among many other attendants of the central star. The beginning of the new era in natural science is marked by Newton's theory of gravitation, which holds the solar system together, and which remained unchallenged for some two centuries. In our time, however, it has been dethroned by EINSTEIN's relativistic theory of gravitation, which completely does away both with the heliocentric system of planets and with gravity acting at a distance.

The position is rather similar in optics, with its change in ideas concerning the nature of light, imagined either as a stream of small particles, according to NEWTON, or as a train of waves in the light-ether, according to HUYGENS. At the beginning of the nineteenth century there occurred the sudden change from the corpuscular theory to the wave theory; the present century, in turn, brought with it a fresh transformation, of which I shall speak presently. In the study of electricity and magnetism, the middle of the last century was a time of revolution, in which the concept of action at a distance was compelled to give place to the idea of a continuous transmission of force through the ether. The profound problem of the structure of matter, which chemistry—a mighty branch of the tree of physical sciences—has made its especial concern, exhibited even a few decades ago the immemorial antithesis of atomistics and the continuum concept. This antithesis today seems to be resolved in favour of the former; yet these problems are bound up with one of the most fundamental revolutions of ideas, which is taking place before our eyes under the name of the quantum theory.

In a smaller scale the rise, acceptance and fall of theories is an everyday occurrence; what today is valuable knowledge will tomorrow be so much junk, hardly worth a historical backward glance. The question thus arises: what then is the value of theories? Are they not perhaps a mere by-product of research, a kind of metaphysical ornament, draped like a lustrous cloak over the 'facts' which alone signify, at best a support and aid in our labour, stimulants to the imagination in conceiving new experiments?

The fact that this question can be proposed at all shows that the meaning of physical theories is by no means obvious, and this is why I have taken that subject as the theme of my lecture today. There are many physicists at the present time, when once again a grave crisis regarding the fundamental ideas of physics has just been overcome, who are not entirely clear what to think of this latest change of theory.

These theories—relativity and the quantum theory—which are characteristic of the present time, are also the best suited for our purposes, since we ourselves feel many of their assertions to be strange, paradoxical, or even meaningless. The older theories must have had a similar effect on their contemporaries; we, however, can conjure up this state of mind only artificially, by historical investigation. As I have paid little attention to the study of history, I shall content myself with a brief glance backward to earlier periods of crisis.

Any theoretical concept originates from observation and its most plausible interpretation. The sight of the fixed, unshakable earth on which we are borne, and of the moving heavens, leads naturally to the geocentric system of the Universe. The fact that light throws sharp shadows can be most simply understood in terms of the corpuscular hypothesis, which is found already, in poetical form, in the works of LUCRETIUS. Of mechanics, which later became a model for all physical theories, antiquity knew only statics, the science of equilibrium. The reason is, of course, that the forces acting upon levers and other machines can be replaced by forces exerted by the human (or animal) body, and thus belong to the realm of things directly perceptible to the senses.

What now is the significance of the change, when these primitive ideas—the geocentric system of the Universe, the corpuscular hypothesis of light, the statical force in mechanics—are replaced by others? The deciding factor is certainly Man's need to believe in a real external world, independent of him and permanent, and his ability to mistrust his sensations in order to maintain this belief. A very distant object seems smaller than when it is near, but Man sees always the 'object,' imagines it to be always the same size, and believes with absolute certainty that he could go and convince himself of the fact by touching and feeling the object. The objects with which primitive Man deals—stones, trees, hills, houses, animals, men—have the property of meeting this test. Such is the origin of geometry, which in its beginnings was entirely the study of the mutual positions and size relations of rigid bodies. In this sense geometry is the most ancient branch of physics; it first showed that objects in the external world follow strict laws as regards their spatial properties. Later, delight in the beauty of these laws had the result that the empirical foundations of geometrical science were disregarded or even denied, and the study of its logical framework became an end in itself, as being a part of mathematics. The geodesist and the astronomer, however, have always regarded the teachings of geometry as statements concerning the real objects in the world, and have never doubted that even bodies which, because of their remoteness, are not directly accessible to us follow the same laws. The application of the rules of geometry to the planets showed that they must be very distant and very large, that their motions on the night sky are only the projections of their true paths in space; and finally the analysis of these paths and the refinement of observational technique led of necessity to the COPERNICAN system. The latter's victory proves that belief in well-tried laws is stronger than a direct sense-impression. Of course, the new theory must explain the reason for this sense-impression, on which the previously accepted doctrine, now recognized to be false, was based. In COPERNICUS' case, it sufficed to point out the size of the earth in comparison with Man. This astronomical example is typical of all subsequent cases. In the stellar universe we have for the first time a reality accessible to only one sense, that of sight, and then often as an insignificant-seeming impression, far removed from the lives and struggles of men, and yet undoubtedly just as real as the chair in which I am sitting or the piece of paper from which I am reading. This objective reality of which I speak is always and everywhere founded on the same principle: obedience to the general laws of geometry and physics. Even the chair I regard as real only because it exhibits the constant properties appertaining to solid bodies of its kind; the geometry and mechanics needed here is at everyone's command from unconscious experience. There is no essential difference when we consider the reasons why we think the point of light called Mars to be a gigantic sphere like the earth; in this case, however, the observations must be more exact and geometry and mechanics must be consciously applied. The simple and unscientific man's belief in reality is fundamentally the same as that of the scientist. Some philosophers concede this standpoint as being practically indispensable for the scientist; it goes under the name of empirical realism and has a precarious position among the various kinds of idealism. Here, however, we do not wish to discuss the quarrels of the different schools of philosophy, but only to state as clearly as possible the nature of the reality which forms the subject of natural science. It is not the reality of sense-perceptions, of sensations, feelings, ideas, or in short the subjective and therefore absolute reality of experience. It is the reality of things, of objects, which form the substratum underlying perception. We take as a criterion of this reality not any one sense-impression or isolated experience, but only the accordance with general laws which we detect in phenomena.

What we have here expounded, using the example of astronomy, occurs over and over again in the development of physics. We have already ascertained in essentials the meaning of all theories, and now wish to show that all the revolutions which have taken place in physics are stages on the road to the construction of an objective world, which combines the macrocosmos of the stars, the microcosmos of the atoms and the cosmos of everyday things into a consistent whole.

Let us first consider mechanics. In its period of simplicity it was, as we have remarked, unable to progress beyond the study of equilibrium. The study of motion or dynamics was the product of a more sophisticated age. The laws which GALILEO and NEWTON derived from their observations cannot be enunciated without ideas which lie far outside the natural limits of thought. Words like *mass* and *force* had, of course, been used earlier: 'mass' meant roughly the amount of some material, 'force' the magnitude of an exertion. In mechanics, however, these words acquire a new precise meaning; they are artificial words, perhaps the first to be coined. Their sound is the same as words of ordinary speech, but their meaning can be found only from a specially formulated definition. I will not discuss this (by no means simple) definition here, but merely mention that a concept occurs therein which, in the days before science, played no part and can indeed be exactly explained only with the aid of mathematical tools, namely the concept of acceleration. If mass is defined by means of this concept (as 'resistance to acceleration'), we already see clearly the foundation of mechanics as an artificial product of the mind. Experience of terrestrial bodies which could be adduced to support the new theory in the period between GALILEO and NEWTON was fairly limited. Yet the inner logic of Galilean mechanics was so strong that NEWTON was able to take the great step of applying it to the motions of the stars. The immense success of this step rests essentially on the idea that the force which the heavenly bodies exert on one another is fundamentally the same as the gravity which we know on earth. This idea, however, caused the abandonment of a concept until then generally accepted, namely that forces from a body are exerted only on its immediate neighbourhood. Only such contact forces were known to statics. Terrestrial gravity, in the work of GALILEO, at first appears only as a mathematical aid in formulating the laws of falling. NEWTON himself regarded the distant action from star to star, which he needed to explain the motions of the planets, only as a provisional hypothesis, to be later replaced by a contact or near-action. The effect of the practical successes of NEWTON's theory of gravitation on his successors was so overwhelming, however, that the distant action of gravitation was not only taken for granted, but was used as a model for the manner of action of other forces, those of electricity and magnetism. Fierce battles have been fought in former times over this distant action across empty space. Some called it a monstrosity opposed to the natural idea of force; others hailed it as a marvellous tool for unlocking the secrets of the stellar universe. Who was right? We say: the Newtonian force of gravitation is an artificial concept, which has little more than its name in common with the simple idea, the feeling of force. Its justification rests merely on its place in the system of objective natural science. So long as it fulfills its duty there, it can remain; but as soon as new observations contradict it, it must give way for the formation of new ideas, which will be required to agree with the distant-action theory within the realm of the older observations. This change has occurred only in our time, after a long development, which was closely connected with the evolution of the sciences of electricity and magnetism.

As we have already said, the forces of electricity and magnetism were, at the time of their first systematic investigation about 150 years ago, interpreted as distant actions on the model of gravitation. COULOMB's law of the attraction of electric charges, BIOT and SAVART's law of the effect of a current on a magnetic pole, are imitations of NEWTON's laws in form and conception. In the mathematical construction of the theory, however, a notable thing occurred: the so-called potential theory was found to give transformations of these laws which put them in the form of near actions, of forces exerted on one another by adjoining points in space. Yet this remarkable equivalence of such heterogeneous concepts went almost unnoticed. New discoveries had to be made in order to compel a physical decision of the question 'distant or near action?' The discoverer of these new facts was FARADAY, and their interpreter was MAXWELL. MAXWELL's equations are a near-action theory of electromagnetic phenomena, and thus signify conceptually a return to a mode of thought closer to the natural mode. I think, however, that this is quite unimportant. What then is the state of affairs? If we exclude FARADAY's and MAXWELL's new discoveries, magnetic induction and dielectric displacement current, MAXWELL's equations contain nothing more than the already existing potential theory, the mathematical transformation of distant-action laws into near-action ones.

The change in physical theories occurring in the middle of the last century is thus, from this viewpoint, not really a revolution, destroying what exists, but a conquest of new territory, involving a reorganization of the old territory.

As a result of this conquest, however, a new concept comes to the fore, that of the universal ether. For every near-action requires a carrier, a substratum between whose particles the forces act, and since the electric and magnetic forces can be transmitted even through empty space, where no ordinary bodies are present, there was nothing for it but to assume an artificial body. This, however, was the easier inasmuch as such an ether had already been invented in another field, that of optics, and the new theory of electricity was in a position to identify this light-ether forthwith with the electromagnetic ether.

We now come to the point where we can glance at the theory of light. Here, as has already been remarked, the issue between the corpuscular and wave theories had been decided in favour of the latter at the beginning of the nineteenth century. Far-reaching as this decision was, it signified, in the same sense as above, more a conquest of new territory with consequent change of government than a true revolution. For, so long as the phenomena of interference and diffraction remained unknown, the concepts of corpuscles and of very short waves were in actual fact equivalent, so that the dispute could not be resolved. The fact that the whole of the eighteenth century adhered to the corpuscular theory was really an accident. Firstly, there was the authority of NEWTON, who had preferred the corpuscular theory as being a simpler concept, in the absence of cogent counter-proofs. Secondly, there existed no mathematical proof that, even with short waves, the occurrence of apparently sharp shadows can be explained; this proof was first furnished by FRESNEL in trying to explain the actual diffuseness of shadows, that is, the phenomena of diffraction. As soon as these phenomena were discussed, there could no longer be any doubt that the wave concept is the correct one. I should like to emphasize that this is still true today, although, as we shall see presently, the corpuscular theory has had a revival. Just as we observe water waves and can follow their propagation, so we can detect light waves with our apparatus. It would be entirely irrational to employ different words and viewpoints in the two cases. This certainty of the existence of light waves leads to the problematical features of the most recent optical discoveries, which we shall discuss below.

First of all, however, we must make a few remarks concerning the *ether problem*. Waves require a carrier, and so it was assumed that space is filled with the light-ether. The first period of the ether theory again showed the simple carrying over of familiar viewpoints. Elastic bodies were known to propagate waves, and so it was assumed that the ether had the same properties as an ordinary elastic substance. It could not, indeed, resemble a gas or a liquid, since only longitudinal waves are propagated in the latter, whereas experiments with polarized light show that light waves are certainly transverse. It was thus necessary to assume a solid elastic ether throughout the Universe, through which light waves are propagated. It is obvious that this gives rise to difficulties when we try to understand why the planets and the other heavenly bodies move through this substance with no noticeable retardation. Nor was it possible to explain satisfactorily the processes of reflection and refraction at surfaces, propagation of light in crystals, and such like. It was thus a relief when MAXWELL's theory was experimentally confirmed by HERTZ, since it was now possible to equate the electromagnetic ether with the light-ether. The formal difficulties disappeared immediately, since the electromagnetic ether is not a mechanical body with properties known from ordinary experience, but an entity of a special kind, with its own laws—like MAXWELL's equations, a typical artificial concept.

The period in physics following MAXWELL was so packed with successes gained by this theory that the belief was often held that all the essential laws of the inorganic world had been discovered. For it proved possible to fit mechanics also into the 'electromagnetic world picture,' as it was called; the resistance to acceleration, caused by the mass, was ascribed to electromagnetic induction effects. Yet the limits of this realm were at hand, visible to the far-seeing, and beyond those limits lay new territory which could not be mastered by the means at hand. With this we enter the most recent period. Its characteristic is that physical criticism takes in ideas which no longer belong exclusively to its province, but are claimed by philosophy as its own. Here, however, we shall always place the physical viewpoints in the foreground.

As always, the conceptual difficulty came upon the theory of the electromagnetic universal ether by a refinement in observational technique: I refer to the celebrated MICHELSON-MORLEY experiment. Before this, the ether could be imagined as a substance at rest everywhere in the Universe, having particular properties, and LORENTZ was able to show that all the then known electromagnetic processes in bodies at rest or in motion could be explained in this way. The real difficulty was to explain the fact that no ether wind can be detected on the earth, which moves at a considerable speed through the ether. LORENTZ was able to show that any optical and electromagnetic effects caused by this ether wind must be extremely small; they are proportional to the square of the ratio of the earth's velocity to that of light, a quantity of the order of 10^{-8}. Such small quantities were below the limit of observability, until MICHELSON's experiment was made. This should therefore have revealed the blowing aside of light waves by the ether wind. It is well known, however, that, like all later repetitions of the experiment, it showed no trace of the effect. This was very difficult to explain, and very artificial assumptions became necessary, such as the hypothesis put forward by FITZ-GERALD and LORENTZ that

all bodies are shortened in the direction of their motion. The riddle was solved by
EINSTEIN in his 'special' theory of relativity, and the salient point in this was a
criticism of the idea of time.

What is time? To the physicist it is not the feeling of elapsing, not the symbol
of becoming and ceasing to be, but a measurable property of processes, like many
others. In the naïve period of science, direct observation or perception of the passage
of time naturally determines the formation of the concept of time, and the one-to-
one correspondence between the passage of time and the content of experience
naturally led to the view that time is the same here and everywhere else in the
Universe. EINSTEIN was the first to question whether this statement has any content
that can be tested empirically. He showed that the simultaneity of events at different
places can be ascertained only if an assumption is made concerning the velocity of
the signals used, and this, in conjunction with the negative result of the ether–wind
experiment, led him to a new definition of simultaneity, which involved a relativization of the concept of time. Two events at different places are not in themselves
simultaneous; they may be so for one observer, but not for another who is in motion
relative to the first observer. The physical concept of space was also caught up in
this change in ideas, especially when EINSTEIN, some years later, revealed the
relation of gravitation to the new conception of space and time. I cannot enter into
this 'general' theory of relativity within the limits of this lecture; I will merely say
that, in the theory of gravitation, it signifies a transition from distant-action to near-
action, and thus an approach to intuitive ideas. On the other hand, it demands a
great step into the abstract: space and time lose all the simple properties which
before then had made geometry and motion theory such convenient tools for
physics. The familiar geometry of EUCLID and the corresponding time are now
reduced to mere approximations to reality; but at the same time it becomes unintelligible why humanity has so far obtained such good results with this approximation. Even today, one obtains satisfactory results with it almost always in practice;
in fact, it is an unfortunate thing that the deviations capable of testing EINSTEIN's
theory are very rare and difficult to observe. Together with the internal consistency
and logicality of the theory, however, they are enough to gain it acceptance from
physicists, apart from a few dissenters.

What is the position regarding the universal ether in the theory of relativity?
EINSTEIN at first proposed to avoid this concept altogether. For the ether might be
thought of as a substance having at least the most elementary properties in common
with ordinary substances. These properties include the recognizability and identi-
fiability of individual particles. In the theory of relativity, however, it is meaningless to say, 'I have been at this point of the ether before.' The ether would be a substance whose parts have neither position nor velocity. Nevertheless, EINSTEIN

later preferred to continue to use the word 'ether,' as a purely artificial concept, of course, having hardly anything in common with the ordinary idea of a substance. For it is simply a grammatical necessity, in speaking of oscillations and waves in space, to have a subject to govern the verb 'to oscillate.' We therefore say, 'The ether oscillates, and does so according to the field equations of EINSTEIN's theory'; and that is all we *can* say about it.

The theory of relativity also modified importantly the concept of mechanical mass, fusing it with that of energy. These are consequences which are of the greatest significance in physics, in connection with investigations of the structure of matter and radiation; they have not, however, aroused so much excitement as the criticism of the traditional ideas of space and time, since the latter were regarded as belonging to the content of philosophy. The fact of the matter is—as it is agreed by all sensible philosophers—that philosophy in former times, when the individual sciences had not detached themselves, merely took over and retained the conceptions of natural science. Since these conceptions, as always in the naïve period, corresponded entirely to sense perception, many schools of thought formed the prejudice that they were an immutable property of the mind, experience *a priori*. This is, of course, true in the realm of perception, but not for the objective realm of physics, whose properties must always be fitted to the progress of experience and its systematic arrangement.

Much though the theory of relativity has brought in the way of innovations, it is yet rather the climax of a development—the doctrine of the continuous universal ether—than the inception of a new period. A new period, however, does begin with the present century by the introduction of PLANCK's quantum theory. Its real and deepest root is in atomistics, an ancient doctrine going back to the Greek philosophers. Before 1900 it had developed quite continuously and peacefully, though more and more richly and fruitfully. Chemistry first made useful the concept of atoms; gradually it conquered physics as well, mainly by explaining the properties of gases and solutions, and from there penetrated into the theory of electricity. The passage of electricity through electrolytic solutions led to the hypothesis of atoms of electricity, called electrons, and these were so brilliantly established in discharge phenomena in gases, and in cathode rays and Becquerel rays, that the reality of the electrons soon became as certain as that of the material atoms. Now, when the electron had been revealed as a kind of sub-atom, investigation was concentrated on the problem of decomposing ordinary atoms into their electric component parts. The idea was that all atoms are built up of electrically negative electrons and of electrically positive components whose nature was not yet known. The difficulty is that, according to simple mathematical theorems, charged bodies can never be at rest in stable equilibrium under the known action of electric forces. It was thus necessary to assume hypothetical unknown forces, and this is, of course, rather unsatisfactory. Then came RUTHERFORD's great discovery. He bombarded atoms with atomic fragments, called α-rays, emitted by radioactive bodies; these rays, by virtue of their very high velocity, penetrate into the interior of the atoms they strike. RUTHERFORD concluded with complete certainty from the deflections undergone by the α-rays that they move as if a heavy and very small positively charged mass, the 'nucleus,' lay at the center of atom, this mass exerting the ordinary electric forces on the α-particles. It thus became in the highest degree improbable that the atom was held together by unknown non-electric forces. But how could the electrons be in equilibrium around the nucleus? The only way out seemed to be to assume that the electrons are not at rest, but move in orbits round the nucleus, like the planets round the Sun. This, of course, did not help much, since such a dynamical system is highly unstable. There is no doubt that our planetary system would be reduced to chaos if it were so unfortunate as to pass close to another large star; yet the atoms of a gas survive a hundred million collisions every second, without the slightest change in their properties.

This astonishing stability of atoms was an utter riddle from the standpoint of the theory as it was at the end of the nineteenth century, nowadays usually called the 'classical theory' for short. An equally difficult puzzle was posed by the gigantic array of facts which the spectroscopists had meanwhile assembled. Here one had a direct message from the interior of the atom, in the form of light oscillations emitted by it, and this message did not sound at all like gibberish, but rather like an orderly language—except that it was unintelligible. For the gases, in particular, a simple structure of the spectrum was recognizable: it consists of individual colored lines, each corresponding to a single periodic oscillation, and these lines exhibited simple regularities. They can be arranged in series in such a way that, from the serial number of the line, its position in the spectrum can be calculated, with the greatest accuracy, from a simple formula. This was first found by BALMER for hydrogen, and later for many other substances by other investigators, in particular RUNGE and RYDBERG. The attractive work of photographing and measuring spectra appealed to a great number of physicists, and so an immense quantity of observational material was accumulated over the years, from which many important conclusions could be drawn concerning individual problems in physics, chemistry and astronomy, but whose real meaning remained hidden. It was the same situation as with the extinct Maya peoples, of whose script numerous specimens have been found in the ruined cities of Yucatan; unfortunately, nobody can read them.

In physics, the key to the riddle was finally discovered, and that by a strangely indirect road. At the turn of the century it was the latest fashion to examine the radiation of glowing solid bodies. Besides the technological importance of the problem in the manufacture of incandescent lamps and so on, profound theoretical results were also hoped for from its solution. For KIRCHHOFF had proved, on the basis of unassailable thermodynamic reasoning, that radiation which leaves the interior of a glowing furnace through a small hole must give a spectrum of an invariable kind, entirely independent of the nature of the substances in the furnace and in its walls; and this conclusion had been confirmed by experiment. From the measurement of 'cavity radiation,' results were therefore expected concerning quite general properties of the process of radiation, and this expectation was not in vain. Nevertheless, it now seems remarkable that one of the most profound laws could be discovered in this way. For—to resume the metaphor of a foreign tongue—one listened not to the articulate words of individuals, but to a crowd shouting all at once, and from this din the key word was heard that made all the others intelligible. The glowing furnace is such a complex structure, containing innumerable oscillating atoms which send out to us their confused assembly of waves. The characteristic feature of the spectrum of this assembly is, by experiment, that it has a definite colour, according to the temperature, red, yellow or white-hot. This means that a certain range of oscillations, depending on the temperature, is most strongly represented, while the intensity gradually falls to zero on both sides of this, towards both rapid and slow oscillations. The classical theory, on the other hand, demanded that the intensity should continually increase on the side of rapid oscillations. Here there was again an insoluble contradiction of the laws accepted at that time.

After countless attempts to ascribe this contradiction to erroneous conclusions within the classical theory had proved abortive, PLANCK in 1900 ventured to propose a positive assertion amounting to this: the energy of the oscillating particles in the furnace alters, not continuously by radiation, but discontinuously, in jumps, and the ratio of the quantum of energy transferred in each jump to the frequency of oscillations in the light emitted or absorbed is a fixed and universal constant. This number, today known as PLANCK's constant, could be quite accurately calculated from experiments then available on heat radiation, and has since been redetermined many times by the most various methods, without any considerable change in the original value.

In fact, a new fundamental constant of nature had been discovered, comparable with the velocity of light or the charge on the electron. This no one doubted, but most people found it very difficult to accept the hypothesis of energy quanta. EINSTEIN alone soon saw that it renders intelligible other peculiarities in the transformation of mechanical energy into radiation. I must say a few words regarding the most important of these phenomena, the so-called photoelectric effect. If light of a given frequency falls on a metal plate in a high vacuum, it is observed that electrons are detached from the plate. The remarkable thing about the process is that only the number, and not the velocity, of the electrons emitted depends on the intensity of the light. The wave picture is of no use in understanding this; for, if we move the metal plate away from the light source, the incident wave becomes weaker and more and more rarefied, and it is incomprehensible how it can always communicate the same energy to an electron. EINSTEIN observed that this behaviour can be immediately understood if the light does not consist of waves, but is a shower of particles; the hail of bullets from a machine-gun thins out with distance, but each individual bullet retains its penetrating power. Combining this idea with PLANCK's quantum hypothesis, EINSTEIN predicted that the energy of the light particle, and therefore that of the ejected electron, must be equal to the frequency multiplied by PLANCK's constant. This result has been entirely confirmed by experiments. Thus we have a revival of the old corpuscular theory of light in a new form.

We shall consider below the conflict arising from this. First, however, let me say a few words regarding the further development of the quantum theory. It is well known that NIELS BOHR conceived the idea of using PLANCK's hypotheses to explain the properties of atoms; he supposed that atoms (quite unlike a classical system of planets) can exist only in a series of discrete states, and that, in a transition from one state to another, light is emitted or absorbed whose frequency is to the energy change of the atom in the ratio given by PLANCK. By this means, all the contradictions mentioned above between experiment and the classical theory are brought back to the same origin, and can be resolved by the assumption of discrete energy quanta. The stability of the atom is explained by the existence of a 'lowest' quantum state, in which the atom remains even when perturbed, provided that the perturbations do not reach the amount of the smallest energy jump possible in the atom. The existence of this lowest energy threshhold was established experimentally by FRANCK and HERTZ, who bombarded atoms (of mercury vapour) with electrons of measured velocity. At the same time, this confirmed BOHR's hypothesis concerning light emission; for as soon as the energy of the bombarding electrons exceeded the first energy threshhold, light of a single colour was emitted, its frequency being that calculated from the energy by means of PLANCK's relation. The whole of the large amount of observational material accumulated by the spectroscopists was thus converted, at one stroke, from a collection of numbers and unintelligible rules to a most invaluable record regarding the possible states of the atoms and the energy differences between them. Further, the previously quite enigmatic conditions for the excitation of the various spectra became completely intelligible.

Despite this enormous success of BOHR's point of view, the road from his simple idea of stationary states to a complete and logically satisfactory mechanics of the atom was a long and laborious one. Here again we have the primitive period, in which the laws of ordinary mechanics were applied as far as possible to the electron orbits in the atom, and it is remarkable that this was in fact possible to some extent, despite the irreconcilable antithesis between the continuous nature of the classical quantities and the discontinuous processes (jumps) of the quantum theory. Finally, however, the necessary modification of mechanics was effected, so as to take account of the discontinuities. The new *quantum mechanics* was evolved in different forms, partly from a fundamental idea due to HEISENBERG, by one group, here in Göttingen and by DIRAC in Cambridge, partly as the so-called wave mechanics of DE BROGLIE and SCHRÖDINGER. These formalisms finally proved to be essentially identical; together, they form a logically closed system, the equal of classical mechanics in internal completeness and external applicability. At first, however, they were only formalisms, and it was a matter of discovering their meaning *a posteriori*. It is, in fact, very common in physical investigations to find it easier to derive a formal relation from extensive observational material than to understand its real significance. The reason for this lies deep in the nature of physical experience: the world of physical objects lies outside the realm of the senses and of observation, which only border on it; and it is difficult to illuminate the interior of an extensive region from its boundaries. In the quantum theory, there were especial difficulties, of which I should like to discuss the most important, namely the revival of the corpuscular theory of light. The idea of individually moving light quanta was supported by a number of further tests, and in particular by COMPTON's experiment. This showed that, when such a light quantum collides with an electron (realized as the scattering of X-rays by substances, such as paraffin, with many loosely-bound electrons) the usual collision laws of mechanics hold, as for billiard balls. The primary light quantum gives up some energy to the electron with which it collides, and so the recoiling light quantum has less energy, and—by PLANCK's relation—a smaller frequency than the primary one. The consequent decrease in frequency of the scattered X-ray has been demonstrated experimentally, and so has the existence of recoil electrons.

There is thus no doubt of the correctness of the assertion that light consists of particles. But the other assertion that light consists of waves is just as correct. In discussing the proofs of the wave nature of light, we have seen that, in every phenomenon of interference, we can perceive the light waves as clearly and evidently as water waves or sound waves. The simultaneous existence of corpuscles and waves, however, seems quite irreconcilable. Nevertheless, the theory must solve the problem of reconciling these two ideas, not of course in the realm of observation, but in that of objective physical relations, where the only criterion of existence, a part from freedom from logical contradiction, is agreement between theoretical predictions and experiment. The solution of this problem was attained by a criticism of fundamental concepts, very similar to that in the theory of relativity.

The basis of the entire quantum theory is PLANCK's relation between energy and frequency, which are asserted to be proportional. In this 'quantum postulate,' however, there is an absurdity. For the concept of energy clearly refers to a single particle (a light quantum or an electron), that is, to something of small extent; the concept of frequency, however, belongs to a wave, which must necessarily occupy a large region of space, and indeed, strictly speaking, the whole of space: if a segment of a purely periodic wave-train is removed, it is no longer periodic. The equating of the energy of a particle and the frequency of a wave is thus in itself quite irrational. It can, however, be made rational, if a principle is renounced which was previously always accepted in physics, namely, that of determinism. Earlier, it had always been supposed that the photoelectric process, in which an electron is ejected from a metal plate by a light wave, is determined in every detail—that there is meaning in the question, 'When and where is an electron ejected?' Or, what is the same thing, 'Which light quantum, at what point and at what time, takes effect on striking the plate?'

Suppose that we decide to renounce this question, an act which is the easier inasmuch as no experimenter would think of asking it, or answering it, in a particular case. It is, in fact, clearly a purely artificial question; the experimenter is invariably content to find out how many particles appear, and with what energy.

Let us therefore not ask where exactly a particle is, but be satisfied to know that it is in a definite, though fairly large, region of space. The contradiction between the wave and corpuscular theories then disappears. This is most easily seen if we allot to the wave the function of determining the probability that a particle will appear, the energy of the particle being related to the frequencies present in the wave by means of PLANCK's relation. If the region of space considered is large, and the wave-train consequently almost unperturbed and purely periodic, there corresponds to it a precise frequency, and a precisely defined particle energy; but the point where the particles appear in this region of space is quite indefinite. If it is desired to determine the position of the particles more exactly, the region of space in which the process is observed must be diminished; by so doing, however, a segment of the wave is removed, and its purely periodic character is destroyed; such a non-periodic disturbance, nevertheless, can be analyzed into a greater or lesser number of purely periodic oscillations; to each of the various frequencies of this mixture, there then corresponds a different energy of the observed particles. Thus an exact determination of position destroys the determination of the energy, and *vice versa*.

This law of restricted measurability discovered by HEISENBERG has been confirmed in every case. For every extensive quantity (such as determinations of position and time), there is an intensive quantity (such as velocity and energy), such that, the more exactly the one is determined, the less accurately can the other be determined, and it is found that the product of the ranges to within which two such associated quantities are known is exactly PLANCK's constant. That is the true significance of this hitherto mysterious constant of nature; it is the absolute limit of accuracy of all measurements. Only its extreme smallness is responsible for the fact that its existence was not discovered earlier.

From this standpoint it is possible to interpret the formalism of quantum mechanics, in any individual case, so that the relation with the observational concepts of the experimenter is shown, without the possibility of any contradiction.

This, of course, does not happen without the sacrifice of familiar ideas. For example, when we speak of a particle, we are accustomed to imagine its entire path in a concrete manner. We may continue to do so, but we must be careful in drawing conclusions therefrom. For, if such an assumed path is to be tested experimentally, the test itself will in general alter the path, no matter how carefully it is performed. More fundamentally important is the abandonment of determinism, the replacement of a rigorously causal description by a statistical one.

Probability and statistics have already played a certain part in physics, in the case of phenomena involving large numbers (e.g., in the kinetic theory of gases). These methods, however, were usually regarded as emergency devices in cases where our knowledge of details is insufficient. Provided that the position and velocity of all the particles in a closed system were known at some instant, the future evolution of the system would be completely determined, and could be predicted by mere calculation. This corresponds to our experience concerning large bodies. Let us recall the story of William Tell. When Tell, before aiming at the apple, sent a brief prayer to Heaven, he surely prayed for a steady hand and a keen eye, believing that the arrow would then find its way into the apple automatically. In precisely the same way, the physicist supposed that his electron and α-ray bullets would certainly hit any desired atom, provided that he could aim accurately enough, and he did not doubt that this was merely a question of practice, which could be solved better and better as experimental technique progressed. Now, on the contrary, it is asserted that the aiming itself can be only of limited accuracy. If Gessler had ordered Tell to shoot a hydrogen atom from his son's head by means of an α-particle, and given him, instead of a crossbow, the best laboratory instruments in the world, then Tell's skill would have been unavailing; whether he hit or missed would have been a matter of chance.

The impossibility of exactly measuring all the data of a state prevents the predetermination of the further evolution of the system. The causality principle, in its usual formulation, thus becomes devoid of meaning. For if it is in principle impossible to know all the conditions (causes) of a process, it is empty talk to say that every event has a cause. Of course, this opinion will be opposed by those who see in determinism an essential feature of natural science. There are others, however, who hold the contrary opinion that quantum mechanics asserts nothing new as regards the question of determinism; that, even in classical mechanics, determinism is only a fiction and of no practical significance;* that, in reality, despite mechanics, there holds everywhere the principle that the basis of all statistics is small causes, great effects. If, for instance, we consider the atoms of a gas as small spheres, the mean free path between two collisions is, at normal pressure, many thousands of times the diameter of the atom; a very slight deviation in the direction of recoil at one collision will therefore convert a direct hit at the next collision into a miss, and a marked change of direction will be replaced by an undisturbed passage. This is certainly so, but it does not yet reach the heart of the matter. Let us return once more to Tell. What better example could we have of the theorem of a small cause and a great effect than shooting at the apple, where the accuracy of the aim is a matter of life and death? Yet the story is evidently based on the conception of the ideal marksman, who can always make the error of his aim smaller than the most diminutive target—supposing, of course, that no unforeseeable influence, such as the wind, diverts his missile. In exactly the same way, we can imagine an ideal case in classical mechanics; a system completely isolated from external influences and an exactly determined initial state, and there is no reason to suppose that any approximation to this aim is not only difficult but impossible. Quantum mechanics, however, asserts that it may be impossible. This distinction may seem pointless to the practical scientist; the discovery of the existence of an absolute limit of accuracy is, however, of great importance in the logical structure of the theory.

Even if we disregard all philosophical aspects, the contradiction between the corpuscular and wave properties of radiation would be insoluble in physics without this statistical viewpoint. This is where the theory has scored a great success: it predicted on formal grounds that even material rays, of emitted atoms or electrons, must exhibit a wave character in suitable circumstances, and the experimenters have since confirmed this prediction by remarkable interference experiments.

Although the new theory then seems well founded on experiment, it may still be asked whether it cannot in future be made again deterministic by extension or refinement. To this we may reply that it can be proved by exact mathematics that the accepted formalism of quantum mechanics admits of no such addition. If therefore it is desired to retain the hope that determination will some day return, the present theory must be regarded as intrinsically false; certain statements of this theory would have to be disproved by experiment. The determinist must therefore

---

*R. VON MISES, *Probability, Statistics and Truth,* Springer, 1928. Compare the following argument with the article "Is Classical Mechanics in Fact Deterministic?," p. 81 of this collection.

not protest but experiment, if he wishes to convert the adherents of the statistical theory.

Of course many people, on the contrary, welcome the abandonment of determinism in physics. I remember that, at the time of the appearance of the earliest work on the statistical interpretation of quantum mechanics, a gentleman approached me with some occultistic pamphlets, thinking I might be suitable for a conversion to spiritualism. There are also, however, serious observers of scientific evolution, who consider the present turn in physics to be the collapse of one conception of the Universe and the beginning of another, deeper idea of the nature of 'reality.' Physics itself, they claim, admits that there are 'gaps in the sequence of determinateness.' What right has it then to put forward its devices as 'realities'?

In meeting such arguments it is important to demonstrate clearly that the new quantum mechanics is no more and no less revolutionary than any other newly propounded theory. Once again, it is really a conquest of new territory; in the course of this it is found, as on previous occasions, that the old principles are no longer wholly adequate, and must be in part replaced by new ones. But the old ideas still remain as a limiting case, comprising all phenomena for which PLANCK's constant can, on account of its smallness, be neglected in comparison with quantities of the same kind. Thus events in the world of large bodies obey to a high accuracy the old deterministic laws; deviations occur only in the atomic range. If quantum mechanics has any peculiarity, it is that it does not decide between two modes of presentation (corpuscles and waves) which previously were equally possible, but, after the seeming victory of one, reinstates the other and combines both in a higher unity. The necessary sacrifice is the idea of determinism; but this does not mean that rigorous laws of Nature no longer exist. Only the fact that determinism is among the ordinary concepts of philosophy has caused us to regard the new theory as particularly revolutionary.

I hope to have shown that the whole evolution of physical theories, up to their latest form, is governed by a consistent striving, and the object of this striving will be clear from the individual examples given. Let me attempt to express it once more in a somewhat more general form. The world of Man's experience is infinitely rich and manifold, but chaotic and involved with the experiencing being. This being strives to arrange his impressions and to agree with others concerning them. Language and art, with their numerous modes of expression, are such ways of transmission from mind to mind, complete in their way where objects of the sense-world are concerned, but not well suited to the communication of exact ideas concerning the outer world. This marks the beginning of the task of science. From the multitude of experiences it selects a few simple forms, and constructs from them, by thought, an objective world of things. In physics, all 'experience' consists of the activity of constructing apparatus and of reading pointer instruments. Yet the results thereby obtained suffice to re-create the cosmos by thought. At first images are formed which are much influenced by observation; gradually, the conceptions become more and more abstract; old ideas are rejected and replaced by new ones. But, however far the constructed world of things departs from observation, nevertheless it is indissolubly linked at its boundaries to the perceptions of the sense, and there is no statement of the most abstract theory that does not express, ultimately, a relation between observations. That is why each new observation shakes up the entire structure, so that theories seem to rise and fall. This, however, is precisely what charms and attracts the scientist. The creation of his mind would be a melancholy thing, did it not die and come to life once more.

CAUSE, PURPOSE AND ECONOMY
IN NATURAL LAWS
[MINIMUM PRINCIPLES IN PHYSICS]

[A lecture given at the Royal Institution of Great Britain Weekly Evening Meeting, February 10, 1939. First published in *Proc. Roy. Inst.*, Vol. xxx, Part iii (1939).]

Without claiming to be a classical scholar I think that the earliest reference in literature to the problems which I wish to treat tonight is contained in VIRGIL's *Aeneid*, Book I, line 368, in the words 'taurino quantum possent circumdare tergo.'

The story, as told at greater length by the later Greek writer ZOSIAS, is this: Dido, sister of King Pygmalion of the Phoenician city of Tyre, a cruel tyrant who murdered her husband, was compelled to fly with a few followers and landed at the site of the citadel of Carthago. There she opened negotiations with the inhabitants for some land and was offered for her money only as much as she could surround with a bull's hide. But the astute woman cut the bull's hide into narrow strips, joined them end to end, and with this long string encompassed a considerable piece of land, the nucleus of her kingdom. To do this she had evidently to solve a mathematical question—the celebrated *problem of Dido*: to find a closed curve of given circumference having maximum area.

Well, we do not know how she solved it, by trial, by reasoning or by intuition. In any case the correct answer is not difficult to guess, it is the circle. But the mathematical proof of this fact has only been attained by modern mathematical methods.

In saying that the first appearance of this kind of problem in *literature* is that quoted above I am not, of course, suggesting that problems of minima and maxima had never occurred before in the *life of mankind*. In fact nearly every application of reason to a definite practical purpose is more or less an attempt to solve such a problem; to get the greatest effect from a given effort, or, putting it the other way round, to get a desired effect with the smallest effort. We see from this double formulation of the same problem that there is no essential distinction between *maximum* and *minimum*; we can speak shortly of an *extremum* and *extremal* problem. The business man uses the word 'economy' for his endeavour to make the greatest profit out of a given investment, or to make a given profit out of the least investment. The military commander tries to gain a certain strategical position with the minimum loss to his side, and maximum loss to the enemy—a procedure described by the experts by the dubious expression 'economy of life.' These examples show how extremal problems depend on ideas taken from human desires, passions, greeds, hatreds; the ends to be achieved are often utterly unreasonable, but once they are accepted as ends they lead to a strictly rational question, to be answered by logical reasoning and mathematics. Our whole life is just this mixture of sense and non-sense, to attain by rational methods aims of doubtful character. Consider our road system: does it meet the simple requirement of providing the shortest connections between inhabited centers? Certainly not.

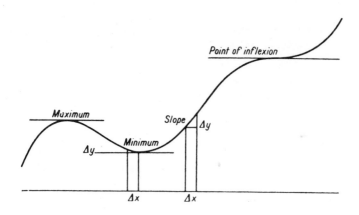

FIG. 1. Maxima, Minima, Points of Inflexion.

The roads are the more or less rational resultant of geographical, historical and economic conditions, which are often anything but rational.

But here we have to do not with the activities of mankind but with the laws of nature. The idea that such laws exist and that they can be formulated in a rational way is a comparatively late fruit of the human intellect. The nations of antiquity developed only a few branches of science, notably geometry and astronomy, both for practical purposes. Geometry arose from the surveying of sites and from archi-tecture, astronomy from the necessities of the calendar and navigation.

Modern science began with the foundation of mechanics by GALILEO and NEW-TON. The distinctive quality of these great thinkers was their ability to free them-selves from the metaphysical traditions of their time and to express the results of observations and experiments in a new mathematical language regardless of any philosophical preconceptions. Although NEWTON was a great theologian his dynamical laws are free from the idea that the individual motion of a planet might bear witness to a definite and detectable purpose. But during his lifetime, at the end of the seventeenth century, geometrical and analytical problems of extremals began to interest mathematicians, and shortly after NEWTON's death in 1727 the meta-physical idea of purpose or economy in nature was linked up with them.

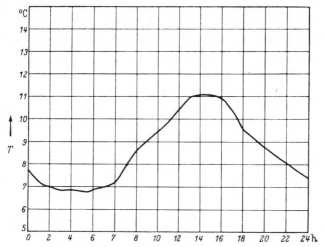

FIG. 2. Diurnal variation of temperature.

Before I go on to speak of the historical development, let us briefly review those geometrical problems exemplified by Dido's land purchase from which we started.

The top of a mountain, the bottom of a valley, are the prototypes of maxima and minima; a vertical profile of a mountain range, as shown in Fig. 1, represents the simplest mathematical figure with extremal points and we see that the tangent line is horizontal at these points. As the figure shows, there are other points with horizontal tangent, but the tangent is a so-called *inflexional tangent*. The common property of these points is that the height is *stationary* in their neighborhood; it does not change appreciably as it would if the point were on a slope.

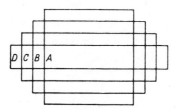

FIG. 3. Rectangles of equal circumference and different area.

	Sides	Circumference	Area
(A)	4 × 4	2 × (4 + 4) = 16	16
(B)	3 × 5	2 × (3 + 5) = 16	15
(C)	2 × 6	2 × (2 + 6) = 16	12
(D)	1 × 7	2 × (1 + 7) = 16	7

You will be acquainted with the method of graphs, representing the law of change of any quantity by a curve on co-ordinate paper. The diurnal variation of temperature, for instance, is shown by a graph like this (Fig. 2); it shows a maximum shortly after noon, and a minimum in the small hours of the morning.

Let us assume that Dido wished to build on her ground a rectangular building with an area as large as possible; this would mean a modification, in fact a great simplification of her problem, as she would not have to choose the curve of maximum area out of all possible closed curves of given length, but merely the rectangle of maximum area out of all rectangles of given circumference. Fig. 3 shows a set of such rectangles which have obviously all a smaller area than the square.

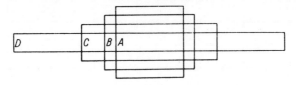

FIG. 4. Rectangles of equal area and different circumference.

	Sides	Area	Circumference
(A)	4 × 4	16	2 × (4 + 4) = 16
(B)	3 × 5.34	16	2 × (3 + 5.34) = 16.7
(C)	2 × 8	16	2 × (2 + 8) = 20
(D)	1 × 16	16	2 × (1 + 16) = 34

This is the simplest form of the genuine *isoperimetric problem* (from the Greek: iso=equal, perimeter=circumference), the general case of which is Dido's problem. But mathematicians nowadays use this name for all kinds of problems in which an extremum has to be determined under a constraining condition (as, for instance, maximum area for given circumference). Here one can generally interchange the two quantities concerned, whereby a maximum of the one becomes a minimum of the other; the square, for instance, is clearly also the rectangle of minimum length surrounding a given area (Fig. 4) and the corresponding fact holds for the circle as compared with all other closed curves.

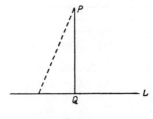

FIG. 5.

Another type of problem is that connected with the idea of the *shortest line*. The simplest case is that of choosing the point Q on a straight line L such that the distance from a given point P outside the line may be as short as possible (Fig. 5). It is evident that Q is the foot of the normal from P to the line L. A little more involved is the question how to find a point Q on a straight line L so that the sum of its distances $P_1Q + QP_2$ from two external points P_1, P_2 is as small as possible.

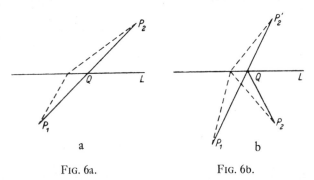

FIG. 6a.   FIG. 6b.

If P_1, P_2 are on different sides of the line L the solution is trivial, namely, Q is the point of intersection of L with the straight line P_1P_2 (Fig. 6a). But if P_1 and P_2 are on the same side of L the solution can easily be found by noticing that to each point P_2 there belongs an 'image' point P_2' on the other side of L, and Q will be the intersection of P_1P_2' with L (Fig. 6b). This idea of an *image* presents the first example of a physical interpretation of such a geometrical problem. For it is evident that if L were a plane mirror a beam of light travelling from P_1 to the mirror and reflected to P_2 would just coincide with our solution. This solution is exactly the optical *law of reflection,* and we have expressed this as a minimum principle: the beam of light selects just that reflecting point Q which makes the total path $P_1Q + QP_2$ as short as possible. I have here a mechanical model to show this: the point Q is represented by a little peg movable along a bar, and the beam of light by a string fixed at one end at P_1, while the other end is in my hand. If I pull the string you see that the point Q adjusts itself so that P_1Q, P_2Q make equal angles with the line, in agreement with the image construction. The light behaves as if each beam had a tendency to contract, and the French philosopher FERMAT has shown that all the laws of geometrical optics can be reduced to the same principle. Light moves like a tired messenger boy who has to reach definite destinations and carefully chooses the shortest way possible. Are we to consider this interpretation as accidental, or are we to see in it a deeper metaphysical significance? Before we can form a judgment we must learn more about the facts and consider other cases.

Let us return to geometrical examples. So far we have assumed that only straight line connections between different points are admitted, or lines composed of straight parts (as in the last example). But this restriction is not necessary, and if it is dropped we approach the domain of problems to which the real Dido problem belongs, namely those where a whole curve has to be determined from the condition that some quantity shall be extremal.

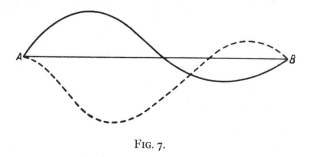

FIG. 7.

The simplest question of this type is: why is *the straight line the shortest connection* between two given points A and B? (Fig. 7). We are here in a much higher branch of mathematics, in the realm of infinite possibilities, called the *calculus of variations*. For we have to compare the length of all possible curves passing through A and B, that is an infinite number of objects which are not points, but figures. It is one of the great triumphs of the human mind to have developed methods for performing this apparently superhuman task.

If we travel on our earth we can never go exactly in a straight line since the earth's surface is not plane. The best we can do is to follow a *great circle,* which is the curve in which the sphere is intersected by a plane passing through the center. Indeed, it can be shown that the shortest path between any two points A, B, not being the ends of the same diameter ('antipodes'), is the arc of the great circle through A and B, or better, the shorter of the two arcs. Ships on the ocean should travel on great circles.

You know that the globe is not an exact sphere but is slightly flattened at the poles, bulging at the equator. What, then, about the shortest line on such a surface?

It is just about a hundred years ago that the great mathematician, KARL FRIED-RICH GAUSS, in Göttingen, hit on this problem when occupied with a geodetic triangulation of his country, the Electorate of Hanover. As he was not merely a surveyor but one of the greatest thinkers of all times, he attacked the problem from the most general standpoint and investigated the shortest lines on arbitrary surfaces. But in remembrance of his starting point, he called these lines *geodesics*. I wish to say a few words about these lines and their properties, as they are in many ways of fundamental importance for physics.

GAUSS' investigation led him to the discovery of non-euclidean geometry. This discovery is generally attributed to the Russian LOBATSCHEFSKY and the Hungarian BOLYAI, and this is quite correct, as these investigators published independently (about 1830) the first systems of non-euclidean geometry. But the discovery (1899) of GAUSS' diary many years after his death and the collection and publication of his correspondence have given ample evidence that a great number of the important mathematical discoveries made by others during the first part of the eighteenth century were already known to him, among them a complete theory of non-euclidean space. He did not publish it because, as he wrote to a friend, he was afraid 'of the clamour of the Boeotians.' This proof that it is possible to construct geometries differing from that of EUCLID without meeting contradictions was a fundamental step towards the modern development of science. It led to an empirical interpretation of geometry as that part of physics which deals with the general properties of the form and position of rigid bodies. Through the work of RIEMANN and EINSTEIN, geometry and physics gradually amalgamated to form a unity. But besides these important developments, the study of geodesics teaches us other things which throw light on the character of different types of physical laws, and on our subject of cause, purpose and economy in nature.

Let us consider a point P on a surface (Fig. 8) and all curves through P which have the same direction at P. It is evident that there is among them a 'straightest curve,' i.e., one with the smallest curvature. I have a model of a surface with the help

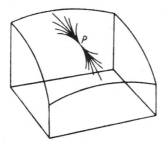

FIG. 8. Lines of minimum curvature on a surface.

of which I can demonstrate to you the straightest curve. There are two small loops fixed on the surface, through which I can thread a piece of a piano wire. This offers resistance to bending in virtue of its elastic properties and, therefore, assumes the straightest shape possible on the surface. I now take a piece of string and pull it through the two loops. This, of course, assumes the shape of the shortest connection between the two points possible on the surface. You see that the straightest line and the shortest line coincide accurately.

Hence the geodesic can be characterized by two somewhat different minimum properties: one which can be called a *local* or *differential property*, namely, to be as little curved as possible at a given point for a given direction; and the other, which can be called *total* or *integral*, namely, to be the shortest path between two points on the surface.

This dualism between 'local' and 'total' laws appears not only here in this simple geometrical problem, but has a much wider application in physics. It lies at the root of the old controversy whether forces act directly at a distance (as assumed in NEWTON's theory of gravitation and the older forms of the electric and magnetic theories), or whether they act only from point to point (as in FARADAY's and MAX-WELL's theory of electromagnetism and all modern field theories). We can illustrate this by interpreting the law of the geodesic itself as a law of physics, in particular of dynamics. NEWTON's first law of dynamics, the *principle of inertia*, states that the straight line is the orbit of any small particle moving free from external forces; a billiard ball moves in a straight line if the table is accurately horizontal so that gravity is inoperative. Imagine a frozen lake so large that the curvature of the earth is perceptible over its length—there is no straight line on it, only straightest lines, the great circles of the globe. It is clear that these are the orbits of free particles. We can, therefore, extend NEWTON's first law to the motion on smooth surfaces by saying that a body free from external forces travels as straight as possible. Here we have a physical law of the *local* character. But, knowing the other minimum property of the geodesic, we can also say: a body always moves from one position to any other by the shortest possible path—which is a law of the *integral* type.

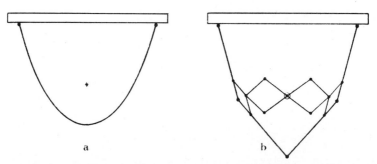

FIG. 9(a). Catenary. (b). Chain of four elements carrying a construction which makes the center of gravity visible.

There seems to be no objection to extremal laws of the local type, but those of the integral type make our modern mind feel uneasy. Although we understand that the particle may choose at a given instant to proceed on the straightest path we cannot see how it can compare quickly all possible motions to a distant position and choose the shortest one—this sounds altogether too metaphysical.

But before we follow out this line of thought we must convince ourselves that minimum properties appear in all parts of physics, and that they are not only correct but very useful and suggestive formulations of physical laws.

One field in which a minimum principle is of unquestionable utility is *statics*, the doctrine of the *equilibrium* of all kinds of systems under any forces. A body moving under gravity on a smooth surface is at rest in stable equilibrium at the lowest points, as this *pendulum* shows. If we have a system composed of different bodies forming a mechanism of any kind, the center of gravity tends to descend as far as possible; to find the configuration of stable equilibrium one has only to look for the minimum of the height of the center of gravity. This height, multiplied by the force of gravity, is called potential energy.

A *chain* [Fig. 9(a)] hanging from both ends assumes a definite shape, which is determined by the condition that the height of the center of gravity is a minimum. If the chain has very many links, we get a curve called the *catenary*. We have here a genuine variational problem of the isoperimetric type, for the catenary has the lowest center among the infinite variety of curves of the same length between the given end-points. I have here a chain consisting of only four links [Fig. 9(b)]. The center of gravity is made visible by a construction of levers (made from light

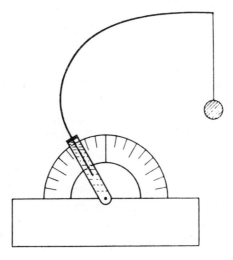

FIG. 10. Steel tape carrying a weight (Elastica).

material so that they do not contribute appreciably to the weight). If I disturb the equilibrium of the chain in an arbitrary way you observe that the center of gravity is always rising.

I will now show you an example where gravity competes with another force, *elasticity* (Fig. 10). I have chosen this special problem, not because it was the subject of my doctor's thesis more than 30 years ago, but because it can be used to explain the difference between the genuine minimum principles of statics and the formal variational principles of dynamics, as we shall see later on. A steel tape is clamped at one end and carries a weight at the other. This weight is pulled downwards by gravity, while the tape tries to resist bending in virtue of its elasticity. This elastic force also has a potential energy; for a definite amount of work must be done to bend the tape into a given curved shape, and it is clear that this energy depends in some way on the curvature of the tape—which varies from point to point. Now there is a definite position of equilibrium, which you see here, namely a position in which the total energy, that of gravitation plus that of elasticity, is as small as possible; if I pull the weight down the gravitational energy decreases, but the energy of elastic bending increases more, so that there results a restoring force; and if I lift the weight, the gravitational energy increases more than the bending energy decreases so that the force is again in the direction towards equilibrium. You see that for some directions of the clamped end there are two positions of equilibrium, one on the left and one on the right.

This also holds for vertical clamping where the two equilibrium forms are symmetrical—but only if the tape is long enough. If I shorten its length sufficiently, the only possible equilibrium form is that in which the tape is straight. There is a definite length for a given weight at which this straight form becomes unstable: it is determined by the condition that beyond this length the potential energy ceases to be a minimum for the straight form and becomes a minimum for a curved form.

The formula for this characteristic length was found by EULER and plays an important role in engineering, as it determines the strength of vertical bars and columns. But similar instabilities also occur for inclined directions of clamping. If I fix the length and change the clamping angle, a jump suddenly occurs from one position to the one on the opposite side. This instability is again determined by the condition of minimum energy. We can summarize the facts connected with the limits of stability by drawing a graph, not of the elastic lines themselves (which are beautiful curves like those shown in Fig. 11, called *elastica*), but by plotting the angle of inclination against the distance from the free end. We now obtain waveshaped curves (Fig. 12), all starting horizontally from the line representing the end carrying the weight. You see that these curves have an envelope and the calculation shows that this envelope is just the limit of stability. Through any point on the right of the envelope there pass at least two curves; this corresponds to the fact that this point represents a clamping angle for which two equilibria exist. If we now move vertically upwards in the diagram, we change the angle of clamping (without changing the length of the tape); when we cross the envelope we pass into a region where there is only one curve through each point. At the envelope one of the configurations becomes unstable and jumps across to the other one. In particular EULER's limit for the stability of the straight form of the tape is represented by the sharp point of the envelope; the distance of this from the origin is just a quarter of the wave length of the neighbor curve, which value gives exactly EULER's formula. I am going to ask you to keep this example in mind as we shall return to it later, when we discuss the minimum principles of dynamics.

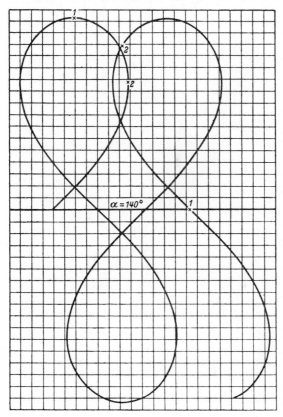

FIG. 11. Elastica.

Another example of the statical principle of minimum energy is provided by *soap bubbles*. Soap films have the property of contracting as much as possible; the potential energy is proportional to the surface-area. A well known experiment shows this very clearly. I project a soap film stretched over a wire in the form of a circle on which a fine thread is fixed. If I destroy the film on one side of the thread, the film on the other side contracts, the thread is pulled tight and assumes the form of an arc of a circle. I now take a closed loop of thread; if I destroy the film in the inner portion the loop immediately forms a perfect circle under the stress of the outer film, showing that this film is under a uniform tension.

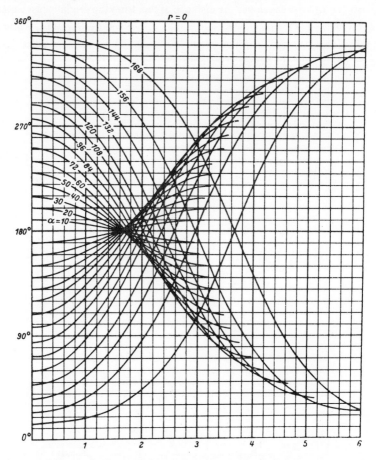

FIG. 12. Diagram showing the limits of stability of the elastica.

It is clear, therefore, that a closed soap bubble filled with air and floating freely in space has the shape of a sphere, which is the minimum surface for a given volume— the spatial analogue of Dido's problem.

There exist other minimum surfaces not closed but determined by a given boundary. We have only to bend a wire to the shape of this boundary and to dip it into a soap solution to get a perfect physical model of the minimum surface. These experiments and their theory were studied long ago by the blind French physicist,

PLATEAU, and you will find a wonderful account of them in the celebrated little book by C. V. BOYS on *Soap Bubbles*. I will show you some of them. See how expert a mathematician Nature is and how quickly she finds the solution!

Some of you may consider these experiments merely as pretty toys without any serious background. But they are chosen only for the sake of illustration. The real importance of the principle of minimum energy can scarcely be exaggerated. All engineering constructions are based on it, and also all structural problems in physics and chemistry.

As an example, I shall show you here some models of *crystal lattices*. A crystal is a regular arrangement of atoms of definite kinds in space. The discovery of LAUE, FRIEDRICH and KNIPPING that X-rays are diffracted by these atomic lattices was used by Sir WILLIAM BRAGG and his son, Professor W. L. BRAGG, for the empirical determination of the atomic arrangements. A great number of these are now well known; for instance, here are two simple models, each consisting of two kinds of atoms in equal numbers per unit of the lattice, but different in structure. One is the lattice of a salt, sodium chloride (NaCl), the other of a similar salt, caesium chloride (CsCl). The question arises, why are they different? The answer can be expected only from a knowledge of the forces between the atoms; for it is clear that the structure is determined by the condition of minimum potential energy. Conversely, a study of this equilibrium condition must teach us something about the character of the atomic forces. I have devoted considerable energy to research in this field; it could be shown that the forces in all these salt crystals are mainly the electrostatic interactions between the atoms which are charged, but that the difference of stability between the two lattice types has its origin in another force, namely the universal cohesion which causes gases to condense at low temperatures. This force, called VAN DER WAALS' attraction, is larger for bigger atoms; and as the caesium atoms are much larger than the sodium atoms the minimum of potential energy is attained for different configurations in caesium and sodium salt.

Considerations of this kind, more or less quantitative, enable us to understand a great number of facts about the internal structure of solid matter.

Similar methods can also be applied to the equilibrium of atoms in molecules, but I shall not discuss them, for the problem of atomic structure is really not one of statics but dynamics, as it involves the motion of electrons in the atom.

Before we proceed to the consideration of minimum principles in dynamics where the situation is not as clear and satisfactory as in statics, we must first mention another part of physics which in a sense occupies an intermediate position between statics and dynamics. It is the theory of heat, *thermodynamics and statistical mechanics*. The phenomena considered are of this type. Substances of different composition and temperature are brought into contact or mixed and the resultant system observed. We have, therefore, to do with the transition from one state of equilibrium to another, but we are not so much interested in the process itself as in the final result. I have here a glass of water and a bottle containing a dye; now I pour the red dye into the water and observe the resultant solution. If we look for a mechanical process with which to compare these processes the nearest is, I think, the elastic steel tape carrying a weight which we have already considered. If one end is fixed vertically there are two stable equilibria; the system can be made to jump over from one to the other by imparting energy to it, but you see that it jumps back again. The process is reversible, it leads to a definite final equilibrium only if the superfluous energy is taken away. But in such a case as that of the mixture of two liquids a final equilibrium is automatically reached and the process is *irreversible*. Not only does it never return spontaneously to the unmixed condition, but even the artificial separation of the dye from the water cannot be performed by any simple means.

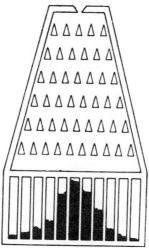

FIG. 13. Galton's quincunx.
(By courtesy of the Institution of Electrical Engineers)

There is a very important extremum principle, discovered by Lord KELVIN, which governs irreversible processes: a certain quantity called *entropy* increases in the process and has a maximum for the final equilibrium state. It is not easy to describe this miraculous entropy in terms of directly observable quantities, such as volume, pressure, temperature, concentration, heat. But its meaning is immediately obvious from the standpoint of atomic theory. What happens if the red solution spreads in the pure water? The molecules of the red dye, at first concentrated in a restricted volume, spread out over a greater volume. A state with a higher degree of order is replaced by one of less order. To explain this expression I have here a model, a flat box, like a little billiard table, into which I can put marbles (purchased at Woolworth's for sixpence). If I place them carefully in the right-hand half, I have a state of partial order; if I shake the box they spread out over the whole box and attain a configuration of lower order. If I throw 20 marbles into the box one after the other so that their position is purely accidental it is very improbable that they will all fall in the right-hand half. One can easily calculate the *probability* of a uniform distribution over the whole box as compared with one in which the majority of the marbles is in the right half; and one finds over-whelming odds in favour of the uniform distribution. Now the statistical theory of heat interprets the entropy of a system with the aid of the probability of the distribution of the atoms, and this helps us to understand why entropy always increases and tends to a maximum.

To show you the working of probability I have here a machine (Fig. 13), invented by GALTON and called the *quincunx*. Shot falls from a hole in the center of the upper end and strikes numerous obstacles in the shape of narrow triangles. At each en-counter the probabilities of falling to the right and to the left are equal. It is clear that a ball has very little chance of always being deflected in the same direction; therefore the cells collecting the balls at the bottom will be comparatively empty at the end, and fuller in the middle. The middle cell corresponds to those balls which have been deflected an equal number of times to the right and to the left, that is to the uniform distribution of deflections. You see that there is a clear maximum. This demonstrates the uniform distribution of the marbles or of the red dye molecules.

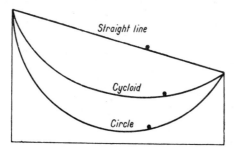

FIG. 14. Brachistochrone.

The thermodynamical principle of maximum entropy is, therefore, really a statistical law and has very little to do with dynamics at all. If a system is initially in a state of partial order, that is, in a state which is not the most probable one (which would correspond to the middle cell of the quincunx) it is very probable that after a while it will have approached the state of maximum probability—or maximum entropy. Very probable, indeed—but not absolutely certain. And the modern technique of micro-observations has revealed cases where deviations from the most probable state are detectable. The extremal principle of statistical mechanics is, therefore, somewhat different in character from the similar laws of pure mechanics. But I cannot go more deeply into the difficult questions of the role of chance and probability in science.

Let us now came back to the *minimum principles of dynamics*.

The first problem of this kind—first both in historical order and in order of simplicity—was formulated at the end of the seventeenth century by JOHANN BERNOULLI of Basle, one of a great family which produced many famous scholars and especially many mathematicians. It is the problem of the curve of quickest descent or *brachistochrone* (Greek: brachys=short, chronos=time): given two points at different levels, not in the same vertical, to determine a connecting curve in such a way that the time taken by a body to slide without friction under the action of gravity from the higher point to the lower is a minimum—compared, of course, with all possible curves through the two points. I have here a model illustrating this problem, but instead of an infinite number of curves I have only three, a straight line, an arc of a circle, and an intermediate curve (Fig. 14). Instead of the bodies sliding without friction I use steel ballbearings rolling on two rails. This has the advantage not only of diminishing friction, but also of retarding the whole motion, which would be too fast without this precaution. As the distance between the rails is a fraction of the diameter of the sphere, it advances for each full rotation only a fraction of the distance it would advance if rolling on a smooth surface. The effect of this trick is only to increase the inertia without changing the gravitational force; the laws of motion are unchanged, only the time scale is reduced.

Now, before I start a race between three balls I ask you to bet if you like which ball will win, and I am prepared to act as bookmaker. It is, of course, not any actual virtue of the ball to be the fastest but of the shape of the curve on which it is rolling.

You see that it is not the straight line which carries the winner, nor the steep descent of the circular arc, but just the intermediate curve. If you were to try with any other curves you would always find the same result; for this curve has been constructed according to the theoretical calculation. It is a so-called *cycloid*, a curve which you can observe hundreds of times every day on the road. It is the curve traced out by a point on the circumference of a wheel rolling along a straight line; I have here a circular disc with a piece of chalk attached to it and if I roll it along the blackboard you see the chalk drawing this line.

The determination of this brachistochronic property of the cycloid was a very satisfactory piece of mathematics; it is a genuine minimum problem and its solution was a great achievement. It attracted much attention and there is no philosopher of this period who did not test his analytical powers by solving similar extremal problems. Another member of the Bernoulli family, DANIEL BERNOULLI, developed during the beginning of the eighteenth century the minimum principle of statics which we have already treated, and applied it to the catenary and the elastic line. Encouraged by these successes DANIEL BERNOULLI raised the question whether it was possible to characterize the orbit, and even the motion in the orbit, of a body subject to given forces—for example, a planet—by a minimum property of the real motion as compared with all other imagined or virtual motions. He put this question to the foremost mathematician of his time, LEONARD EULER, who was very much interested in it and spent several years in investigating it. In the autumn of 1743 he found a solution which he explained with the help of various examples in an appendix to a book on isoperimetric problems published in 1744. It is the basis of the *principle of least action* which has played so prominent a part in physics right up to the present time. But the history of this principle is an amazing tangle of controversies, quarrels over priority and other unpleasant things. MAUPERTUIS, in the same year, 1744, presented a paper to the Paris Academy in which he substituted for FERMAT's optical principle of the shortest light path, which we have already discussed, a rather arbitrary hypothesis and extended the latter, in 1746, to all kinds of motions. He defined *action*, following LEIBNIZ, as the product of mass into the velocity and the distance travelled, and he put forward the universal principle that this quantity is a minimum for the actual motion. He never gave a satisfactory proof of his principle (which is not surprising as it is incorrect) but defended it by metaphysical arguments based on the economy of nature. He was violently attacked, by CHEVALIER D'ARCY in Paris, SAMUEL KÖNIG from Bern and others who showed that if MAUPERTUIS' principle were true, thrifty nature would be forced in certain circumstances to spend not a minimum but a maximum of action. EULER, whose principle is quite correct, behaved rather strangely; he did not claim his own rights but even expressed his admiration for MAUPERTUIS' principle which he declared to be more general. The reasons for this attitude are difficult to trace. One of them seems to be the publication by KÖNIG of a fragment of an alleged letter of LEIBNIZ in which the principle was enunciated. The genuineness of this letter could never be proved and it seems probable that it was a forgery designed to weaken MAUPERTUIS' position. This may have brought EULER over to the side of MAUPERTUIS who was at this time President of the Berlin Academy and a special favourite of the KING FREDERIC II, later known as the Great. The dispute was now carried over into the sphere of the court of Sanssouci and even into the arena of politics. VOLTAIRE, friend of Frederic, who heartily disliked the haughty President of the Academy, took the side of the 'underdog,' KÖNIG, and wrote a caustic pamphlet, 'Dr. Akakia,' against

MAUPERTUIS. But the King, although he thoroughly enjoyed VOLTAIRE's witty satire, could not sacrifice his grand President and was compelled to defend MAUPERTUIS. This led at last to the disruption of their friendship and to VOLTAIRE's flight from Berlin, as described in many biographies of Frederic and of VOLTAIRE.

The curse of confusion has rested for a long period on the principle of least action. LAGRANGE, whose work was the culmination of the development of NEWTON's dynamics, gives an unsatisfactory formulation of the principle. JACOBI restricts it in such a way that the minimum condition determines the orbit correctly; the motion in the orbit must be found with the help of the energy equation. This was an important step. But the spell was at last broken by the great Irishman, Sir WILLIAM ROWAN HAMILTON, whose principle is mathematically absolutely correct, simple and general. At the same time it put an end to the interpretation of the principle expressing the economy of nature. Let us look quite briefly at the real situation.

We have already considered the quantity called *potential energy* of a set of forces, which is the amount of work which must be done to bring the mechanical system into a given configuration and therefore represents a measure of the ability of the system to do work. This potential energy depends only on the configuration and has its minimum in the equilibrium position. If the system is in motion part of the potential energy is converted into energy of motion or *kinetic energy*, namely the sum of half the mass into the square of the velocity of the particles. The law of conservation of energy states that the sum of the two forms of energy is always constant. Now the principle of Hamilton has to do not with the sum but with the difference of these two kinds of energy. It states that the law of motion is such that a quantity frequently called *action*, namely the sum of the contributions of each time interval to the difference of kinetic and potential energy, is stationary for the actual motion, as compared with all virtual motions starting at a given time from a given configuration and arriving at a given subsequent time at another given configuration.

Purposely I say stationary, not minimum, for indeed there is in general no minimum.

What really happens can be explained very clearly with the help of the *simple pendulum*. For there is, by a kind of fortunate mathematical coincidence, a statical problem for which the genuine minimum principle for the potential energy coincides formally with the principle of least action for the pendulum. This is our old friend the *steel tape*. In fact, the sum of the bending energy of the weight attached is exactly the same mathematical expression as the total action of the pendulum (the sum of the contributions of all time elements to the difference of kinetic and potential energy); therefore, the curves representing the angle of inclination of the elastic line as a function of the distance from the free end are exactly the same lines as those representing the angle of deflection of the pendulum as a function of time. You see the vibrational character in the graph although only a small part of the curve is drawn.

Now we have seen that only those regions of the graph, which are simply covered by the lines, correspond to a real minimum, a stable configuration of the elastic line. There are other regions, those beyond the envelope, where two or more lines pass a given point. Only one of those lines corresponds to a real minimum. But both represent possible motions of the pendulum. Although the conditions at the ends of the elastic tape do not correspond exactly to those at the ends of the time interval in HAMILTON's principle, there is this fact in common. If the length of the tape, or the corresponding time interval in HAMILTON's principle for the pendulum exceeds a certain limit, there is more than one possible solution, and not each of them can correspond to a true minimum, though to a possible motion. In this way we come to the conclusion that the actual motion is not in every case distinguished by a genuine extremal property of action, but by the fact that the action is stationary as explained at the beginning of the lecture.

Thus the interpretation in terms of economy breaks down. If nature has a purpose expressed by the principle of least action it is certainly not anything comparable with that of a business man. We may, I think, regard the idea of finding purpose and economy in natural laws as an absurd piece of anthropomorphism, a relic of a time when metaphysical thinking dominated science. Even if we accept the idea that nature is so thrifty with her stock of action that she tried to save it as long as possible—she succeeds, as we have seen, only during the first small part of the motion—we cannot help wondering why she considers just this strange quantity as especially valuable.

The *importance of* HAMILTON's *principle* lies in a different direction altogether.

It is not nature that is economical but science. All our knowledge starts with collecting facts, but proceeds by summarizing numerous facts by simple laws, and these again by more general laws. This process is very obvious in physics. We may recall, for instance, MAXWELL's electromagnetic theory of light by which optics became a branch of general electrodynamics. The minimum principles are a very powerful means to this end of unification. This is easily understood by considering the simplest example, that of the shortest path. If a military commander has a good map he can move his troops from one given point to another by simply announcing the point of destination, without caring much about the details of the route, since he supposes that the officer of the detachment will always march by the shortest route. This minimum principle, together with the map, regulates all possible movements. In the same way the minimum principles of physics replace innumerable special laws and rules—always supposing the map, or in this case the kinetic and potential energy, are given.

The ideal would be to condense all laws into a single law, a *universal formula,* the existence of which was postulated more than a century ago by the great French astronomer LAPLACE.

If we follow the Viennese philosopher, ERNST MACH, we must consider economy of thought as the only justification of science. I do not share this view; I believe that there are many other aspects and justifications of science but I do not deny that economy of thought and condensation of the results are very important, and I consider LAPLACE's universal formula as a legitimate ideal. There is no question that the Hamiltonian principle is the adequate formulation of this tendency. It would be the universal formula if only the correct expressions for the potential energy of all forces were known. Nineteenth century thinkers believed, more or less explicitly, in this program and it was successful in an amazing degree.

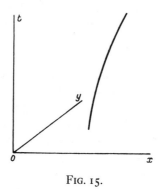

FIG. 15.

By choosing a proper expression for the potential energy nearly all phenomena could be described, including not only the dynamics of rigid and elastic bodies but also that of fluids and gases, as well as electricity and magnetism, together with electronic theory and optics. The culmination of this development was EINSTEIN's *theory of relativity,* by which the abstract principle of least action regained a simple geometrical interpretation, at least that part of it depending on the kinetic energy. For this purpose one has to consider time as a fourth co-ordinate, as Fig. 15 shows (where one dimension of space is omitted); a motion is then represented by a line in this 4-dimensional world in which a non-euclidean geometry is valid, of the type invented by RIEMANN. The length of this line between two points is just the kinetic part of the action in HAMILTON's principle, and the lines representing motions (under the action of gravity) are *geodesics* of the 4-dimensional space. EINSTEIN's law of gravitation, which contains NEWTON's law as a limiting case, can also be derived from an extremum principle in which the quantity which is an extremum can be interpreted as the total curvature of the space-time world. But these are abstract considerations on which I cannot dwell here.

We call this period of physics, which ends with the theory of relativity, the classical period in contrast to the recent period which is dominated by the *quantum theory*.

The study of atoms, their decomposition into nuclei and electrons, and the disintegration of the nuclei themselves has led to the conviction that the laws of classical physics do not hold down to these minute dimensions. A new mechanics has been developed which explains the observed facts very satisfactorily but deviates widely from classical conceptions and methods. It gives up strict determinism and replaces it by a statistical standpoint. Consider as an example the spontaneous disintegration of a radium atom; we cannot predict when it will explode but we can establish exact laws for the probability of the explosion, and therefore predict the average effects of a great number of radium atoms. The new mechanics assumes that all laws of physics are of this statistical character. The fundamental quantity is a *wave function* which obeys laws similar to those of acoustical or optical waves; it is not, however, an observable quantity but determines indirectly the probability of observable processes. The point which interests us here is the fact that even this abstract wave function of quantum mechanics satisfies an extremum principle of the Hamiltonian type.

We are still far from knowing LAPLACE's universal formula but we may be convinced that it will have the form of an extremal principle, not because nature has a will or purpose or economy, but because the mechanism of our thinking has no other way of condensing a complicated structure of laws into a short expression.

APPENDIX

As the argument against the economic interpretation of the principle of least action rests on the comparison of the dynamical problem of the pendulum and the statical problem of the loaded elastic tape, readers who know some mathematics may welcome a few formulae showing the identity of the variational principles for these two examples.

If l is the length of the string, and θ is the angle of deflection, (Fig. 16a) then $\frac{d\theta}{dt}$ is the angular and $\frac{l d\theta}{dt}$ the linear velocity; therefore the kinetic energy $T = \frac{1}{2} m l^2 \left(\frac{d\theta}{dt} \right)^2$ where m is the mass of the bob. The height of the bob above its lowest position is, as the figure shows, $l - l \cos \theta$. Multiplying this by the weight mg (g acceleration of gravity) we get the potential energy; but as a constant does not matter we can omit mgl and write the potential energy $U = - mgl \cos \theta$. The difference of kinetic and potential energy is $T - U = \frac{1}{2}ml^2\left(\dfrac{d\theta}{dt}\right)^2 + mgl \cos \theta$, and the action during the time interval from $t = 0$ to $t = \tau$ is $\displaystyle\int_0^\tau \left\{\frac{1}{2}A\left(\dfrac{d\theta}{dt}\right)^2 + W \cos \theta\right\} dt$, where the abbreviations $A = ml^2$ and $W = mgl$ are used.

We now consider the elastic tape. The energy stored up in the element ds of the

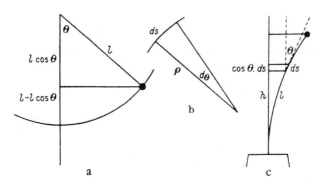

FIG. 16.

tape by bending it into a curve of radius of curvature ρ is $\dfrac{1}{2}A\dfrac{1}{\rho^2}ds$, where A is the bending modulus. The figure (16b) shows that $ds = \rho d\theta$, so that the bending energy of ds is $\dfrac{1}{2}A\left(\dfrac{d\theta}{ds}\right)^2 ds$, and the total elastic energy of bending $\displaystyle\int_0^l \frac{1}{2}A\left(\dfrac{d\theta}{ds}\right)^2 ds$, where l is the length of the tape.

The potential energy of the weight W attached to the end is Wh where h is the height of this weight above the level of the clamped end. The figure (16c) shows that h consists of contributions $\cos \theta\, ds$ of the single elements of the tape; therefore the potential energy of the weight is $\displaystyle\int_0^l W \cos \theta\, ds$. Adding these two potential energies we get

$$\int_0^l \left(\frac{1}{2}A\left(\frac{d\theta}{ds}\right)^2 + W \cos . \theta\right) ds,$$

an expression which is identical with the action of the pendulum if the element ds and the total length l of the tape are replaced by the time element dt and the total time τ in the case of the pendulum.

EINSTEIN'S STATISTICAL THEORIES

[First published in 'The Library of Living Philosophers,' *Albert Einstein: Philosopher-Scientist,* Vol. vii (1949).]

One of the most remarkable volumes in the whole of scientific literature seems to me Vol. 17 (4th series) of *Annalen der Physik*, 1905. It contains three papers by EINSTEIN, each dealing with a different subject, and each today acknowledged to be a masterpiece, the source of a new branch of physics. These three subjects, in order of pages, are: theory of photons, Brownian motion, and relativity.

Relativity is the last one, and this shows that EINSTEIN's mind at that time was not completely absorbed by his ideas on space and time, simultaneity and electrodynamics. In my opinion he would be one of the greatest theoretical physicists of all times even if he had not written a single line on relativity—an assumption for which I have to apologize, as it is rather absurd. For EINSTEIN's conception of the physical world cannot be divided into watertight compartments, and it is impossible to imagine that he should have bypassed one of the fundamental problems of the time.

Here I purpose to discuss EINSTEIN's contributions to statistical methods in physics. His publications on this subject can be divided into two groups: an early set of papers deals with classical statistical mechanics, whereas the rest is connected with quantum theory. Both groups are intimately connected with EINSTEIN's philosophy of science. He has seen more clearly than anyone before him the statistical background of the laws of physics, and he was a pioneer in the struggle for conquering the wilderness of quantum phenomena. Yet later, when out of his own work a synthesis of statistical and quantum principles emerged which seemed to be acceptable to almost all physicists, he kept himself aloof and sceptical. Many of us regard this as a tragedy—for him, as he gropes his way in loneliness, and for us who miss our leader and standard-bearer. I shall not try to suggest a resolution of this discord. We have to accept the fact that even in physics fundamental convictions are prior to reasoning, as in all other human activities. It is my task to give an account of EINSTEIN's work and to discuss it from my own philosophical standpoint.

EINSTEIN's first paper of 1902, 'Kinetische Theorie des Wärmegleichgewichtes und des zweiten Hauptsatzes der Thermodynamik'[1] is a remarkable example of the fact that when the time is ripe important ideas are developed almost simultaneously by different men at distant places. EINSTEIN says in his introduction that nobody has yet succeeded in deriving the conditions of thermal equilibrium and of the second law of thermodynamics from probability considerations, although MAXWELL and BOLTZMANN came near to it. WILLARD GIBBS is not mentioned. In fact, EINSTEIN's paper is a re-discovery of all essential features of statistical mechanics and obviously written in total ignorance of the fact that the whole matter had been thoroughly treated by GIBBS a year before (1901). The similarity is quite amazing. Like GIBBS, EINSTEIN investigates the statistical behaviour of a virtual assembly of equal mechanical systems of a very general type. A state of the single system is described by a set of generalized co-ordinates and velocities, which can be represented as a point in a $2n$-dimensional 'phase-space'; the energy is given as function of these variables. The only consequence of the dynamical laws used is the theorem of LIOUVILLE according to which any domain in the $2n$-dimensional phase-space of all co-ordinates and momenta preserves its volume in time. This law makes it possible to define regions of equal weight and to apply the laws of probability. In fact, EINSTEIN's method is essentially identical with GIBBS' theory of canonical assemblies. In a second paper, of the following year, entitled 'Eine Theorie der Grundlagen der Thermodynamik,'[2] EINSTEIN builds the theory on another basis not used by GIBBS, namely on the consideration of a single system in course of time (later called 'Zeit-Gesamtheit,' time assembly), and proves that this is equivalent to a certain virtual assembly of many systems, GIBBS' micro-canonical assembly. Finally, he shows that the canonical and micro-canonical distribution lead to the same physical consequences.

EINSTEIN's approach to the subject seems to me slightly less abstract than that of GIBBS. This is also confirmed by the fact that GIBBS made no striking application of his new method, while EINSTEIN at once proceeded to apply his theorems to a case of utmost importance, namely to systems of a size suited for demonstrating the reality of molecules and the correctness of the kinetic theory of matter.

This was the theory of Brownian movement. EINSTEIN's papers on this subject are now easily accessible in a little volume edited and supplied with notes by R. FÜRTH, and translated into English by A. D. COWPER.[3] In the first paper (1905) he sets out to show 'that according to the molecular-kinetic theory of heat, bodies of microscopically visible size suspended in a liquid will perform movements of such magnitude that they can be easily observed in a microscope, on account of the molecular motion of heat,' and he adds that these movements are possibly identical with the 'Brownian motion' though his information about the latter is too vague to form a definite judgment.

The fundamental step taken by EINSTEIN was the idea of raising the kinetic theory of matter from a possible, plausible, useful hypothesis to a matter of observation, by pointing out cases where the molecular motion and its statistical character can be made visible. It was the first example of a phenomenon of thermal fluctuations, and his method is the classical paradigm for the treatment of all of them. He regards the movement of the suspended particles as a process of diffusion under the action of osmotic pressure and other forces, among which friction due to the viscosity of the liquid is the most important one. The logical clue to the understanding of the phenomenon consists in the statement that the actual velocity of the suspended particle, produced by the impacts of the molecules of the liquid on it, is unobservable; the visible effect in a finite interval of time τ consists of irregular displacements, the probability of which satisfies a differential equation of the same type as the equation of diffusion. The diffusion coefficient is nothing but the mean square of the displacement divided by 2τ. In this way EINSTEIN obtained his celebrated law expressing the mean square displacement for τ in terms of measurable quantities (temperature, radius of the particle, viscosity of the liquid) and of the number of molecules in a gram-molecule (AVOGADRO's number N). By its simplicity and clarity this paper is a classic of our science.

In the second paper (1906) EINSTEIN refers to the work of SIDENTOPF (Jena) and GOUY (Lyon) who convinced themselves by observations that the Brownian motion was in fact caused by the thermal agitation of the molecules of the liquid, and from this moment on he takes it for granted that the 'irregular motion of suspended particles' predicted by him is identical with the Brownian motion. This and the following publications are devoted to the working out of details (e.g., rotatory Brownian motion) and presenting the theory in other forms; but they contain nothing essentially new.

I think that these investigations of EINSTEIN have done more than any other work to convince physicists of the reality of atoms and molecules, of the kinetic theory of heat, and of the fundamental part of probability in the natural laws. Reading these papers one is inclined to believe that at that time the statistical aspect of physics was preponderant in EINSTEIN's mind; yet at the same time he worked on relativity where rigorous causality reigns. His conviction seems always to have been, and still is today, that the ultimate laws of nature are causal and deterministic, that probability is used to cover our ignorance if we have to do with numerous particles, and that only the vastness of this ignorance pushes statistics into the forefront.

Most physicists do not share this view today, and the reason for this is the development of quantum theory. EINSTEIN's contribution to this development is great. His first paper of 1905, mentioned already, is usually quoted for the interpretation of the photo-electric effect and similar phenomena (STOKES' law of photo-luminescence, photo-ionization) in terms of light-quanta (light-darts, photons). As a matter of fact, the main argument of EINSTEIN is again of a statistical nature, and the phenomena just mentioned are used in the end for confirmation. This statistical reasoning is very characteristic of EINSTEIN, and produces the impression that for him the laws of probability are central and more important by far than any other law. He starts with the fundamental difference between an ideal gas and a cavity filled with radiation: the gas consists of a finite number of particles, while radiation is described by a set of functions in space, hence by an infinite number of variables. This is the root of the difficulty of explaining the law of black body radiation; the monochromatic density of radiation turns out to be proportional to the absolute temperature (later known as the law of RAYLEIGH-JEANS) with a factor independent of frequency, and therefore the total density becomes infinite. In order to avoid this, PLANCK (1900) had introduced the hypothesis that radiation consists of quanta of finite size. EINSTEIN, however, does not use PLANCK's radiation law, but the simpler law of WIEN, which is the limiting case for low radiation density, expecting rightly that here the corpuscular character of the radiation will be more evident. He shows how one can obtain the entropy S of black body radiation from a given radiation law (monochromatic density as function of frequency) and applies then BOLTZMANN's fundamental relation between entropy S and thermodynamic probability W,

$$S = k \log W$$

where k is the gas constant per molecule, for determining W. This formula was certainly meant by BOLTZMANN to express the physical quantity S in terms of the combinatory quantity W, obtained by counting all possible configurations of the atomistic elements of the statistical ensemble. EINSTEIN inverts this process: he starts from the known function S in order to obtain an expression for the probability which can be used as a clue to the interpretation of the statistical elements. (The same trick has been applied by him later in his work on fluctuations;[4] although this is of considerable practical importance, I shall only mention it, since it introduces no new fundamental concept apart from that 'inversion.')

Substituting the entropy derived from WIEN's law into BOLTZMANN's formula, EINSTEIN obtains for the probability of finding the total energy E by chance compressed in a fraction αV of the total volume V

$$W = \alpha^{E/h\nu};$$

that means, the radiation behaves as if it consisted of independent quanta of energy of size $h\nu$ and number $n = E/h\nu$. It is obvious from the text of the paper that this result had an overwhelming power of conviction for EINSTEIN, and that it led him to search for confirmation of a more direct kind. This he found in the physical phenomena mentioned above (e.g., photoelectric effect) whose common feature is the exchange of energy between an electron and light. The impression produced on the experimentalists by these discoveries was very great. For the facts were known to many, but not correlated. At that time EINSTEIN's gift for divining such correlations was almost uncanny. It was based on a thorough knowledge of experimental facts combined with a profound understanding of the present state of theory, which enabled him to see at once where something strange was happening. His work at that period was essentially empirical in method, though directed to building up a consistent theory—in contrast to his later work when he was more and more led by philosophical and mathematical ideas.

A second example of the application of this method is the work on specific heat.[5] It started again with a theoretical consideration of that type which provided the strongest evidence in EINSTEIN's mind, namely on statistics. He remarks that PLANCK's radiation formula can be understood by giving up the continuous distribution of statistical weight in the phase-space which is a consequence of LIOUVILLE's theorem of dynamics; instead, for vibrating systems of the kind used as absorbers and emitters in the theory of radiation most states have a vanishing statistical weight and only a selected number (whose energies are multiples of a quantum) have finite weights.

Now if this is so, the quantum is not a feature of radiation but of general physical statistics, and should therefore appear in other phenomena where vibrators are involved. This argument was obviously the moving force in EINSTEIN's mind, and it became fertile by his knowledge of facts and his unfailing judgment of their bearing on the problem. I wonder whether he knew that there were solid elements for which the specific heat per mole was lower than its normal value 5.94 calories, given by the law of DULONG-PETIT, or whether he first had the theory and then scanned the tables to find examples. The law of DULONG-PETIT is a direct consequence of the law of equipartition of classical statistical mechanics, which states that each co-ordinate or momentum contributing a quadratic term to the energy should carry the same average energy, namely $\frac{1}{2} RT$ per mole where R is the gas constant; as R is a little less than 2 calories per degree and an oscillator has 3 co-ordinates and 3 momenta, the energy of one mole of a solid element per degree of temperature should be $6 \times \frac{1}{2}R$, or 5.94 calories. If there are substances for which the experimental value is essentially lower, as it actually is for carbon (diamond), boron, silicon, one has a contradiction between facts and classical theory. Another such contradiction is provided by some substances with poly-atomic molecules. DRUDE had proved by optical experiments that the atoms in these molecules were performing oscillations about each other; hence the number of vibrating units per molecule should be higher than 6 and therefore the specific heat higher than the DULONG-PETIT value—but that is not always the case. Moreover EINSTEIN could not help wondering about the contribution of the electrons to the specific heat. At that time vibrating electrons in the atom were assumed for explaining the ultra-violet absorption; they apparently did not contribute to the specific heat, in contradiction to the equipartition law.

All these difficulties were at once swept away by EINSTEIN's suggestion that the atomic oscillators do not follow the equipartition law, but the same law which leads to PLANCK's radiation formula. Then the mean energy would not be proportional to the absolute temperature but decrease more quickly with falling temperature in a way which still depends on the frequencies of the oscillators. High frequency oscillators like the electrons would at ordinary temperature contribute nothing to the specific heat, atoms only if they were not too light and not too strongly bound. EINSTEIN confirmed that these conditions were satisfied for the cases of poly-atomic molecules for which DRUDE had estimated the frequencies, and he showed that the measurements of the specific heat of diamond agreed fairly well with his calculation.

But this is not the place to enter into a discussion of the physical details of EINSTEIN's discovery. The consequences with regard to the principles of scientific knowledge were far-reaching. It was now proved that the quantum effects were not a specific property of radiation but a general feature of physical systems. The old rule 'natura non facit saltus' was disproved: there are fundamental discontinuities, quanta of energy, not only in radiation but in ordinary matter.

In EINSTEIN's model of a molecule or a solid these quanta are still closely connected with the motion of single vibrating particles. But soon it became clear that a considerable generalization was necessary. The atoms in molecules and crystals are not independent but coupled by strong forces. Therefore the motion of an individual particle is not that of a single harmonic oscillator, but the super-position of many harmonic vibrations. The carrier of a simple harmonic motion is nothing material at all; it is the abstract 'normal mode,' well known from ordinary mechanics. For crystals in particular each normal mode is a standing wave. The introduction of this idea opened the way to a quantitative theory of thermodynamics of molecules and crystals and demonstrated the abstract character of the new quantum physics which began to emerge from this work. It became clear that the laws of micro-physics differed fundamentally from those of matter in bulk. Nobody has done more to elucidate this than EINSTEIN. I cannot report all his contributions, but shall confine myself to two outstanding investigations which paved the way for the new micro-mechanics which physics at large has accepted today—while EINSTEIN himself stands aloof, critical, sceptical, and hoping that this episode may pass by and physics return to classical principles.

The first of these two investigations has again to do with the law of radiation and statistics.[6] There are two ways of tackling problems of statistical equilibrium. The first is a direct one, which one may call the combinatory method: after having established the weights of elementary cases one calculates the number of combinations of these elements which correspond to an observable state; this number is the statistical probability W, from which all physical properties can be obtained (e.g., the entropy by BOLTZMANN's formula). The second method consists in determining the rates of all competing elementary processes, which lead to the equilibrium in question. This is, of course, much more difficult; for it demands not only the counting of equally probable cases but a real knowledge of the mechanism involved. But, on the other hand, it carries much further, providing not only the conditions of equilibrium but also of the time-rate of processes starting from non-equilibrium configurations. A classical example of this second method is BOLTZMANN's and MAXWELL's formulation of the kinetic theory of gases; here the elementary mechanism is given by binary encounters of molecules, the rate of which is proportional to the number-density of both partners. From the 'collision equation' the distribution function of the molecules can be determined not only in statistical equilibrium, but also for the case of motion in bulk, flow of heat, diffusion, etc. Another example is the law of mass-action in chemistry, established by GULDBERG and WAAGE; here again the elementary mechanism is provided by multiple collisions of groups of molecules which combine, split, or exchange atoms at a rate proportional to the number-density of the partners. A special case of these elementary processes is the monatomic reaction, where the molecules of one type spontaneously explode with a rate proportional to their number-density. This case has a tremendous importance in nuclear physics: it is the law of radio-active decay. Whereas in the few examples of ordinary chemistry, where monatomic reaction has been observed, a dependence of reaction velocity on the physical conditions (e.g., temperature) could be assumed or even observed, this was not the case for radio-activity: the decay constant seemed to be an invariable property of the nucleus, unchangeable by any external influences. Each individual nucleus explodes at an unpredictable moment; yet if a great number of nuclei are observed, the average rate of disintegration is proportional to the total number present. It looks as if the law of causality is put out of action for these processes.

Now what EINSTEIN did was to show that PLANCK's law of radiation can just be reduced to processes of a similar type, of a more or less non-causal character. Consider two stationary states of an atom, say the lowest state 1 and an excited state 2. EINSTEIN assumes that if an atom is found to be in the state 2 it has a certain probability of returning to the ground state 1, emitting a photon of a frequency which, according to the quantum law, corresponds to the energy difference between the two states; i.e., in a big assembly of such atoms the number of atoms in state 2 returning to the ground state 1 per unit time is proportional to their initial number—

exactly as for radio-active disintegration. The radiation, on the other hand, produces a certain probability for the reverse process $1 \to 2$ which represents absorption of a photon of frequency ν_{12} and is proportional to the radiation density for the frequency.

Now these two processes alone balancing one another would not lead to PLANCK's formula; EINSTEIN is compelled to introduce a third one, namely an influence of the radiation on the emission process $2 \to 1$, 'induced emission,' which again has a probability proportional to the radiation density for ν_{12}.

This extremely simple argument together with the most elementary principle of BOLTZMANN's statistics leads at once to PLANCK's formula without any specification of the magnitude of the transition probabilities. EINSTEIN has connected it with a consideration of the transfer of momentum between atom and radiation, showing that the mechanism proposed by him is not consistent with the classical idea of spherical waves but only with a dart-like behaviour of the quanta. Here we are not concerned with this side of EINSTEIN's work, but with its bearing on his attitude to the fundamental question of causal and statistical laws in physics. From this point of view this paper is of particular interest. For it meant a decisive step in the direction of non-causal, indeterministic reasoning. Of course, I am sure that EINSTEIN himself was—and is still—convinced that there are structural properties in the excited atom which determine the exact moment of emission, and that probability is called in only because of our incomplete knowledge of the pre-history of the atom. Yet the fact remains that he has initiated the spreading of indeterministic statistical reasoning from its original source, radio-activity, into other domains of physics.

Still another feature of EINSTEIN's work must be mentioned which was also of considerable assistance to the formulation of indeterministic physics in quantum mechanics. It is the fact that it follows from the validity of PLANCK's law of radiation that the probabilities of absorption $(1 \to 2)$ and induced emission $(2 \to 1)$ are equal. This was the first indication that interaction of atomic systems always involves two states in a symmetrical way. In classical mechanics an external agent like radiation acts on the definite state, and the result of the action can be calculated from the properties of this state and the external agent. In quantum mechanics each process is a transition between two states which enter symmetrically into the laws of interaction with an external agent. This symmetrical property was one of the deciding clues which led to the formulation of matrix mechanics, the earliest form of modern quantum mechanics. The first indication of this symmetry was provided by EINSTEIN's discovery of the equality of up- and down-ward transition probabilities.

The last of EINSTEIN's investigations which I wish to discuss in this report is his work on the quantum theory of monatomic ideal gases.[7] In this case the original idea was not his but came from an Indian physicist, S. N. BOSE; his paper appeared in a translation by EINSTEIN[8] himself who added a remark that he regarded this work as an important progress. The essential point in BOSE's procedure is that he treats photons like particles of a gas with the method of statistical mechanics but with the difference that these particles are not distinguishable. He does not distribute individual particles over a set of states, but counts the number of states which contain a given number of particles. This combinatory process together with the physical conditions (given number of states and total energy) leads at once to PLANCK's radiation law. EINSTEIN added to this idea the suggestion that the same process ought to be applied to material atoms in order to obtain the quantum theory of a monatomic gas. The deviation from the ordinary gas laws derived from this theory is called 'gas degeneracy.' EINSTEIN's papers appeared just a year before the discovery of quantum mechanics; one of them contains moreover (p. 9 of the second paper) a reference to DE BROGLIE's celebrated thesis, and the remark that a scalar wave field can be associated with a gas. These papers of DE BROGLIE and EINSTEIN stimulated SCHRÖDINGER to develop his wave mechanics, as he himself confessed at the end of his famous paper.[9] It was the same remark of EINSTEIN's which a year or two later formed the link between DE BROGLIE's theory and the experimental discovery of electron diffraction; for, when DAVISSON sent me his results on the strange maxima found in the reflection of electrons by crystals, I remembered EINSTEIN's hint and directed ELSASSER to investigate whether those maxima could be interpreted as interference fringes of DE BROGLIE waves. EINSTEIN is therefore clearly involved in the foundation of wave mechanics, and no alibi can disprove it.

I cannot see how the BOSE-EINSTEIN counting of equally probable cases can be justified without the conceptions of quantum mechanics. There a state of equal particles is described not by noting their individual positions and momenta, but by a symmetric wave function containing the co-ordinates as arguments; this represents clearly only one state and has to be counted once. A group of equal particles even if they are perfectly alike, can still be distributed between two boxes in many ways—you may not be able to distinguish them individually but that does not affect their being individuals. Although arguments of this kind are more metaphysical than physical, the use of a symmetric wave function as representation of a state seems to me preferable. This way of thinking has, moreover, led to the other case of gas degeneracy, discovered by FERMI and DIRAC, where the wave function is skew, and to a host of physical consequences confirmed by experiment.

The BOSE-EINSTEIN statistics was, to my knowledge, EINSTEIN's last decisive positive contribution to physical statistics. His following work in this line, though of great importance by stimulating thought and discussion, was essentially critical. He refused to acknowledge the claim of quantum mechanics to have reconciled the particle and wave aspects of radiation. This claim is based on a complete re-orientation of physical principles: causal laws are replaced by statistical ones, determinism by indeterminism. I have tried to show that EINSTEIN himself has paved the way for this attitude. Yet some principle of his philosophy forbids him to follow it to the end. What is this principle?

EINSTEIN's philosophy is not a system which you can read in a book; you have to take the trouble to abstract it from his papers on physics and from a few more general articles and pamphlets. I have found no definite statement of his about the question 'What is Probability?'; nor has he taken part in the discussions going on about VON MISES' definition and other such endeavours. I suppose he would have dismissed them as metaphysical speculation, or even joked about them. From the beginning he has used probability as a tool for dealing with nature just like any scientific device. He has certainly very strong convictions about the value of these tools. His attitude toward philosophy and epistemology is well described in his obituary article on ERNST MACH:[10]

Nobody who devotes himself to science from other reasons than superficial ones, like ambition, money making, or the pleasure of brain-sport, can neglect the questions, what are the aims of science, how far are its general results true, what is essential and what based on accidental features of the development?

Later in the same article he formulates *his empirical creed* in these words:

Concepts which have been proved to be useful in ordering things easily acquire such an authority over us that we forget their human origin and accept them as invariable. Then they become 'necessities of thought,' 'given *a priori*,' etc. The path of scientific progress is then, by such errors, barred for a long time. It is therefore no useless game if we are insisting on analysing current notions and pointing out on what conditions their justification and usefulness depends, especially how they have grown from the data of experience. In this way their exaggerated authority is broken. They are removed, if they cannot properly legitimate themselves; corrected, if their correspondence to the given things was too negligently established; replaced by others, if a new system can be developed that we prefer for good reasons.

That is the core of the young EINSTEIN, thirty years ago. I am sure the principles of probability were then for him of the same kind as all other concepts used for describing nature, so impressively formulated in the lines above. The EINSTEIN of today is changed. I translate here a passage of a letter from him which I received about four years ago (November 7th, 1944): 'In our scientific expectation we have grown antipodes. You believe in God playing dice and I in perfect laws in the world of things existing as real objects, which I try to grasp in a wildly speculative way.' These speculations distinguish indeed his present work from his earlier writings. But if any man has the right to speculate it is he whose fundamental results stand like a rock. What he is aiming at is a general field-theory which preserves the rigid causality of classical physics and restricts probability to masking our ignorance of the initial conditions or, if you prefer, of the pre-history, of all details of the system considered. This is not the place to argue abut the possibility of achieving this. Yet I wish to make one remark, using EINSTEIN's own picturesque language: if God has made the world a perfect mechanism, he has at least conceded so much to our imperfect intellect that, in order to predict little parts of it, we need not solve innumerable differential equations but can use dice with fair success. That this is so I have learned, with many of my contemporaries, from EINSTEIN himself. I think, this situation has not changed much by the introduction of quantum statistics; it is still we mortals who are playing dice for our little purposes of prognosis—God's actions are as mysterious in classical Brownian motion as in radio-activity and quantum radiation, or in life at large.

EINSTEIN's dislike of modern physics has not only been expressed in general terms, which can be answered in a similarly general and vague way, but also in very substantial papers in which he has formulated objections against definite statements of wave mechanics. The best known paper of this kind is one published in collaboration with PODOLSKY and ROSEN.[11] That it goes very deep into the logical foundations of quantum mechanics is apparent from the reactions it has evoked. NIELS BOHR has answered in detail; SCHRÖDINGER has published his own sceptical views on the interpretation of quantum mechanics; REICHENBACH deals with this problem in the last chapter of his excellent book, *Philosophic Foundations of Quantum Mechanics,* and shows that a complete treatment of the difficulties pointed out by EINSTEIN, PODOLSKY, and ROSEN needs an overhaul of logic itself. He introduces a three-valued logic, in which apart from the truth-values 'true' and 'false,' there is an intermediate one, called 'indeterminate,' or, in other words, he rejects the old principle of *'tertium non datur,'* as has been proposed long before, from purely mathematical reasons, by BROUWER and other mathematicians. I am not a logician, and in such disputes always trust that expert who last talked to me. My attitude to statistics in quantum mechanics is hardly affected by formal logic, and I venture to say that the same holds for EINSTEIN. That his opinion in this matter differs from mine is regrettable, but it is no object of logical dispute between us. It is based on different experience in our work and life. But in spite of this, he remains my beloved master.

REFERENCES

1   A. EINSTEIN, *Annalen der Physik* (1902) (4), **9**, p. 477.
2   A. EINSTEIN, *Annalen der Physik* (1903) (4), **11**, p. 170.
3   A. EINSTEIN, *Investigations on the Theory of the Brownian Movement*. London: Methuen & Co. (1926).
4   A. EINSTEIN, *Annalen der Physik* (1906) (4), **19**, p. 373.
5   A. EINSTEIN, *Annalen der Physik* (1907) (4), **22**, p. 180.
6   A. EINSTEIN, *Phys. Z.* (1917), **18**, p. 121.
7   A. EINSTEIN, *Berl. Ber.* (1924) p. 261; (1925) p. 318.
8   S. N. BOSE (1924) *Zeitschrift fur Physik*, **26**, 178.
9   E. SCHRÖDINGER, *Annalen der Physik* (1926) (4), **70**, p. 361; s.p. 373.
10   A. EINSTEIN, *Phys. Z.* (1916) **17**, p. 101.
11   A. EINSTEIN, B. PODOLSKY and N. ROSEN (1935) *Phys. Rev.*, **47**, 777.

PHYSICS IN THE LAST
FIFTY YEARS*

[First published in *Nature*, Vol. 168, p. 625 (1951).]

The following review is based on personal recollections and cannot claim historical accuracy and completeness. I shall tell you what has impressed me most, since I attended, in 1901, my first lecture at the University of Breslau, my home city. We were taught what is called today classical physics, which was at that time believed to be a satisfactory and almost complete description of the inorganic world. But even MAXWELL's theory of the electromagnetic field was, about 1900, not a part of the ordinary syllabus of a provincial German university, and I remember well the impression of bewilderment, admiration and hope which we received from the first lecture on this subject given to us by the then young and progressive lecturer CLEMENS SCHAEFER (still active at Cologne).

The first great event of a revolutionary character happened in 1905 with EIN-STEIN's theory of relativity. I was at that time in Göttingen and well acquainted with the difficulties and puzzles encountered in the study of electromagnetic and optical phenomena in moving bodies, which we thoroughly discussed in a seminar held by HILBERT and MINKOWSKI. We studied the recent papers by LORENTZ and POIN-CARÉ, we discussed the contraction hypothesis brought forward by LORENTZ and FITZGERALD and we knew the transformations now known under LORENTZ' name. MINKOWSKI was already working on his four-dimensional representation of space and time, published in 1907, which became later the standard method in fundamental physics. Yet EINSTEIN's simple consideration by which he disclosed the epistemo-logical root of the problem (the impossibility of defining absolute simultaneity of distant events because of the finite velocity of light signals) made an enormous impression, and I think it right that the principle of relativity is connected with his name, though LORENTZ and POINCARÉ should not be forgotten.

Although relativity can rightly be regarded as the culmination of nineteenth-century physics, it is also the mainspring of modern physics because it rejected traditional metaphysical axioms, NEWTON's assumption about the nature of space and time, and affirmed the right of the man of science to construct his ideas, in-cluding philosophical concepts, according to the empirical situation. Thus a new

---

* Substance of a paper read on August 13th before Section A (Mathematics and Physics) of the British Association meeting at Edinburgh.

era of physical science began by an act of liberation similar to that which broke the authority of PLATO and ARISTOTLE in the time of the Renaissance.

That result of relativity which later proved to be the most important, namely, the equivalence of mass and energy as expressed by the formula $E = mc^2$, was at that time considered to be of great theoretical, but scarcely of any practical, interest.

In 1913 EINSTEIN's first attempt on general relativity became known; it was perfected two years later. It is the first step not only beyond Newtonian *meta-physics*, but also beyond Newtonian physics. It is based on an elementary but so far unexplained fact—that all bodies fall with the same acceleration. To this day it is this empirical foundation which I regard as the corner-stone of the enormous mathematical structure erected on it. The logical way which led from this fact to the field equations of gravitation seems to me more convincing than even the confirmation of the astronomical predictions of the theory, as the precession of the perihelion of Mercury, the deflection of light by the sun and the gravitational shift of spectral lines.

EINSTEIN's theory led to a revival of cosmology and cosmogony on an unprecedented scale. I am not competent to judge whether it was the theory which stimulated the astronomers to build bigger and more powerful instruments, or whether the results obtained with these, like HUBBLE's discovery of the expanding universe, stimulated the theoreticians to still loftier speculations about the universe. The result, however, is undoubtedly that our astronomical horizon today, in 1951, is vastly wider, our ideas about the creation vastly grander than they were at the beginning of the period. We can estimate the actual age of the world (some thousand millions of years), its present size (determined by the receding nebulæ reaching the velocity of light) and the total number of nebulæ, stars and atoms, and we have good reasons for assuming that the laws of physics are the same throughout this vast expanse. The names of FRIEDMAN, LEMAÎTRE, EDDINGTON and ROBERTSON must here be mentioned.

But after this boast let me conclude this section on a note of modesty. The fundamental problem of connecting gravitation with other physical forces, to explain the strange value of the gravitational constant, is still unsolved in spite of EDDING-TON's obstinate, ingenious attempts. The most promising idea seems to me that of DIRAC, developed by JORDAN, that the gravitational constant is not a constant at all, but a scalar field quantity, which like the other ten, the components of the metric tensor, undergoes a secular change and has acquired its present value in the course of time elapsed since the creation of the universe.

Before speaking about the most characteristic features of modern physics, atomistics and the quantum concept, I have to dwell for a short time upon classical physics which, of course, has not suddenly ceased to exist, but continues and flourishes to such a degree that I should venture to say: by far the greatest part of the time and effort of physicists is still devoted to problems of this kind, even of those, frequently found in the United States, who believe that nuclear research is the only decent pursuit deserving the name of physics.

In fact, the progress and success since 1900 in ordinary mechanics, elasticity, acoustics, hydro- and aero-dynamics, thermodynamics, electrodynamics and optics is spectacular enough. You have only to remember that in 1900 the internal combustion engine was in its infancy, motor-cars often brought in by horses and the aeroplane a fantastic dream. It would be impossible to attempt even the crudest sketch of these and other technical developments due to physics. Let me only mention a few characteristic points.

The first is the adoption of a more realistic attitude. In the nineteenth century the mechanics of solids and fluids were beautiful mathematical theories well suited for providing examination papers. Today, they tackle actual problems of daily life and technology, for example, in hydrodynamics, boundary layers, heat transfer, forces on moving rigid bodies like the wings of aeroplanes, the stability of these, even for supersonic velocities. Among the pioneers whom I personally knew are G. I. TAYLOR, PRANDTL, KÁRMÁN. In elasticity we have a similar development; the narrow field of problems accessible to analytical solutions has been enormously extended by numerical methods (SOUTHWELL's relaxation method) and the results are checked by photoelastic observations on transparent models.

This trend has been strongly assisted by the invention of mechanical and electrical computing machines. The speed and power of the modern instruments based on electronic valves has stirred the imagination of the world and given rise to a new science, cybernetics, the advocates of which expect a revolution of human civilization from these artificial brains—a belief which I do not share.

Acoustics, the branch of elasticity dealing with the propagation of waves, was confronted by numerous problems through the invention of the gramophone, the telephone and broadcasting. Here again the electronic valve was a powerful tool. Ultrasonic vibrations have been used for studying the elastic properties of crystals, for signalling and for timekeeping. The clock controlled by the oscillations of a piezo-quartz crystal seems to be more accurate and reliable than ordinary pendulum clocks.

Prof. ANDRADE has given an account of the origin and the development of thermodynamics which in 1900 was considered to be complete, with its two fundamental theorems (conservation of energy, increase of entropy). But this complacent conviction was wrong here as in many other cases.

In 1907 NERNST added a third theorem concerning the behaviour of substances at zero temperature. Of its numerous applications to physics and physical chemistry I can only mention the prediction of chemical equilibria and reactions, as exemplified by HABER's method of fixing nitrogen from the air (1914). The experimental approach to absolute zero made great strides. KEESOM arrived in 1931 at 0.7°K with the help of liquid helium. GIAUQUE and MacDOUGALL devised in 1933 a new method for cooling, using the demagnetization of paramagnetic salts. The absolute scale of temperature was extended below 1°K by KURTI and SIMON (1938) and others. Strange phenomena were discovered in this region, the supra-conductivity of metals by KAMERLINGH-ONNES in 1911, and the superfluidity of liquid helium by KEESOM and WOLFKE in 1927, ALLEN and MEISNER, KAPITZA and others.

Even at higher temperatures new phenomena were found, for example, in the field of highly concentrated electrolytic solutions where the names of BJERRUM, G. N. LEWIS, DEBYE and HÜCKEL must be mentioned.

An approach to extreme conditions from another angle was made by BRIDGMAN (since 1905), who systematically investigated the properties of matter under high pressure, reaching more than 100,000 atmospheres. His latest triumph is the observation of the breakdown of the electronic shells of alkali atoms under pressure.

Of great importance seem to me the recent investigations started by ONSAGER in 1930 and continued by CASIMIR, PRIGOGINE, DE BOER and DE GROOT, by which thermodynamics is generalized so as to apply to irreversible processes, by combining the classical laws of flow with one single result of statistical mechanics, the so-called principle of microscopic reversibility. The results seem to have a bearing on the understanding of the processes going on in living organisms.

The progress of electrodynamics is obvious to everybody in technical applications: improvements in the production of power and its transmission over long distances; telecommunication methods, such as telegraphy, telephony and wireless transmission. In 1900 electromagnetic waves were a laboratory experiment. Since MARCONI's success in 1895 broadcasting has become a powerful factor in human affairs.

Electromagnetic waves comprise the whole of optics, but it would be quite impossible to give an account of the progress in all branches of optical research and practice. The improvements and refinements of all kinds of optical apparatus, of the experimental and theoretical investigation of diffraction, refraction, absorption and scattering are enormous. Let me mention only a few outstanding achievements in spectroscopy because of their bearing on atomic physics: the discovery of the ZEEMAN and STARK effect, the disentanglement of spectral series by RYDBERG, PASCHEN, RUNGE, RITZ and others, the RAMAN effect, the extension of the spectrum towards the ultraviolet and infra-red, and finally the closing of the gap, still existing in 1900, between the longest light or heat waves and the shortest radio waves. The pressure of war helped to develop the method known as radar. In the laboratory it provided the magnetic resonance effect, used for the study of atoms, molecules, crystals (CLEETON and WILLIAMS, 1934; GRIFFITH, 1948), and even for the determination of nuclear spin and quadrupole moments (RABI, 1938). It has also enriched our knowledge of the world at large by the application to the ionosphere (APPLETON and BARNETT, BREIT and TUVE, 1925) and to celestial bodies. Reflections have been obtained from the moon (U.S. Signal Corps, 1948) and from meteors (HEY and STEWART, 1946), and waves coming from the Milky Way (JANSKI, 1931) have been observed. This new radio-astronomy will have a profound influence on cosmology.

We now come to atomistics. Although firmly established in the nineteenth century, there were still, in 1900, some distinguished physicists who did not believe in atoms. Today, such people would be regarded as 'cranks,' since the evidence for the atomistic structure of matter is overwhelming.

There are two different but closely interwoven problems to be answered by atomistics: (1) What is the nature of the atoms? (2) How can the behaviour of matter in bulk be accounted for in terms of the collective action of atoms?

Let us begin with the latter question, as it has been answered for a special type of matter already in the nineteenth century: I mean the kinetic theory of gases and its extension to more general systems in statistical equilibrium through GIBBS' statistical mechanics. This was in 1900 a reasonable hypothesis. But EINSTEIN's explanation of the Brownian movement in 1904 and SMOLUCHOWSKI's consecutive work in 1906 provided direct physical evidence for the correctness of the kinetic theory and led PERRIN in 1909 to a reliable value of the number of atoms in the gram-molecule.

The theory of compressed gases and condensation started by VAN DER WAALS in 1873 has been much improved and modernized by URSELL (1927), MAYER (1937) and others.

A statistical treatment of paramagnetism was given by LANGEVIN in 1905, and extended to ferro-magnetism by WEISS in 1907. This was the first example of a type of statistical problem dealing with so-called order-disorder phenomena, to which, for example, the properties of alloys belong. These methods are today of great practical importance.

The logical foundations of statistical mechanics were critically examined by PAUL and TATYANA EHRENFEST (1911) and its mathematical methods vastly developed by DARWIN and FOWLER (1922).

While a satisfactory kinetic theory of liquids, in spite of great efforts, is still lacking even to this day, our knowledge of the solid state has been greatly increased. This work is closely connected with research on X-rays. The nature of X-rays was controversial until 1912. Selective absorption and polarization discovered by BARKLA in 1909 indicated wave structure. A year later W. H. BRAGG found evidence for corpuscular structure. In 1912 POHL and WALTER obtained diffraction at a slit from which Sommerfeld estimated the wave-length. The dispute was finally settled in favour of waves when LAUE and his collaborators found, in 1912, diffraction of X-rays by crystals, demonstrating at the same time the atomistic nature of solids, the lattice structure of crystals, which had been hypothetically assumed for a long time.

In the hands of W. H. and W. L. BRAGG this method opened a new science, atomistic crystallography, which abounds in ingenious experiments and mathematical considerations, as the systematic application of group theory initiated by SOHNKE as early as 1879 and perfected by SCHÖNFLIES and FEDOROW in 1891.

Upon this empirical geometry of crystal lattices there has been erected a dynamical theory which actually started as one of the first applications of quantum theory with EINSTEIN's work of 1907 on the specific heat of solids at low temperatures, and its refinements by DEBYE and by KÁRMÁN and myself in 1910, which, however, has also a large field of application in the classical domain, predicting relations between elastic, thermal and optical properties of crystals. While for a time the ideal lattice was the central object of study, we begin today to understand the reasons why actual crystals do in many ways deviate from this ideal pattern.

Many of these investigations are independent of a detailed knowledge of the atoms themselves, using only some crude averages of their geometrical and dynamical properties, like diameter, charge, dipole moment, polarizability, etc.

The problem that remains is to understand these averages; that means, to investigate the nature of the atoms themselves.

The research in the structure of the atom is intimately connected with radioactivity. The discovery of radioactivity belongs to the nineteenth century. Its rapid development is mainly due to one man—Lord RUTHERFORD. He demonstrated the atomistic character of the α- and β-radiation by counting the particles, using first the scintillation method of CROOKES (1903), later the Geiger counter (1908). In the later development of counting methods a decisive factor was the amplifying electronic valve, invented in its simplest form (diode) by FLEMING in 1904 and improved (triode, pentode, etc.) by DE FOREST in 1907 and LANGMUIR in 1915.

Let me mention here some other experimental techniques of great importance which enable us not only to count but also actually to see the tracks of particles: C. T. R. WILSON's cloud chamber (1911) and its refinement by BLACKETT (1937), the counter-controlled cloud chamber. Then the method of photographic tracks discovered by BLAU and WURMBACHER in 1937, which through the improvement of emulsions has become a most efficient tool for studying atomic processes.

The first revolutionary results, obtained with the then available primitive experimental technique by RUTHERFORD and SODDY about 1900, were the laws of radioactive disintegration which shattered the belief in the invariability of the chemical elements. These laws differ from the ordinary deterministic laws of classical physics, being intrinsically statistic and indeterministic.

At the same time ample proof for the existence of isotopes was found among the radioactive elements. Later, in 1913, J. J. THOMSON discovered the first example of isotopy among ordinary elements (neon) by electromagnetic deflection. From here came on one hand ASTON's mass spectrograph (1919), the renewal of PROUT's hypothesis and the modern version of the Periodic Table with its arrangement of the atoms according to nuclear charge (atomic number Z) as opposed to mass (mass number A); on the other hand, the separation of isotopes in bulk as performed today on an industrial scale for the production of fissionable material.

The distinction between these two numbers Z and A is mainly due to RUTHER- FORD's second great discovery (1911), the nucleus, obtained through the observation of scattering of α-rays. The result that COULOMB's law is valid down to nuclear dimensions suggested to RUTHERFORD the planetary model of the atom, with the nucleus in place of the sun, and electrons in place of the planets. A welcome con- firmation of this was soon (1913) provided by MOSELEY with the help of X-ray spectra. But formidable theoretical difficulties arose because of the lack of stability of such systems according to the laws of classical mechanics.

In fact, atomic research had reached here a point where progress was not possible without a radical change of our fundamental conceptions.

This revolution of thought was already in progress. It had started in 1900, just at the beginning of this period of review, when PLANCK convinced himself that the observed spectrum of black bodies could not be accounted for by classical me- chanics, and put forward the strange assumption that finite quanta of energy ε exist which are proportional to the frequency ν, $\varepsilon = h\nu$.

The physical world received this suggestion with great scepticism as it did not fit at all into the well established wave theory of light. Years passed without much happening. But in 1905 EINSTEIN took up PLANCK's idea and gave it a new turn; he showed that by assuming the light to be composed of particles, later called photons, a quantitative explanation of the photoelectric effect in metals and of similar phenomena is obtained. Using EINSTEIN's interpretation MILLIKAN (1910) derived from measurements of the photo-effect a value of h in excellent agreement with PLANCK's original one.

Further evidence for the existence of quanta was given again by EINSTEIN in 1907 through his theory of specific heat, mentioned already, which not only removed some very disquieting paradoxes of the kinetic theory, but also served as a sound foundation of the modern theory of molecules and crystals.

The final triumph of quantum theory was BOHR's application to RUTHERFORD's planetary model in 1913. It solved the riddle of atomic stability, explained the mysterious spectral series and the main features of the periodic system.

BOHR was, right from the beginning, quite clear that the appearance of the quantum meant a new kind of natural philosophy, and so it has turned out. Yet at the same time BOHR was anxious to keep the connection with classical theory as close as possible, which he succeeded in doing with the help of his principle of correspondence.

There followed a period of about twelve years in which BOHR's ideas were confirmed and developed. Here are a few outstanding events:

FRANCK's and HERTZ' experiments to demonstrate the existence of stationary states with the help of electronic collisions (1914). The disentanglement of the multiplet spectra, including X-rays, by numerous authors, theoretically guided by BOHR and SOMMERFELD. LANDÉ's formula for the Zeeman effect (1921) which led finally to the suggestion of the spinning electron by UHLENBECK and GOUDSMIT (1925). The confirmation of SOMMERFELD's 'quantization of direction' by STERN and GERLACH (1921). The refinement of the theory of the periodic system by BOHR himself, confirmed at once by the discovery of one of the missing elements, hafnium, by COSTER and HEVESY (1922). Then—most important—PAULI's exclusion principle (1924), which gave a theoretical foundation to striking features of observation. Finally, the Compton effect (1923), which demonstrated the usefulness of EINSTEIN's conception of photons.

Thus the paradoxical situation had to be faced that both the undulatory and the corpuscular theory of light were right—in fact, PLANCK's formula $\varepsilon = h\nu$ states a relation between these contradictory hypotheses.

This challenge to reason came to a climax through DE BROGLIE's famous thesis of 1924 in which this duality wave-corpuscle was, by a purely theoretical argument, extended to electrons. The first confirmation was given by ELSASSER (1927) with the help of experiments on electron scattering on metals made by DAVISSON and GERMER (1927), and soon these authors, and independently G. P. THOMSON, the son of the discoverer of the electron as a particle, produced diffraction patterns with metal foils which established definitely the existence of DE BROGLIE's waves.

May I mention here in parenthesis that the idea of the electron microscope is considerably older than this theory; it was first suggested in 1922 by H. BUSCH on the grounds of considerations analogous to geometrical optics. After DE BROGLIE the wave theory of optical instruments became applicable and the resolving power could be determined. I cannot dwell on details, but I wish to remind you that today not only bacteria and viruses but even big molecules can be made visible and photographed.

The duality wave-corpuscle made an end of the naïve intuitive method in physics which consists in transferring concepts familiar from everyday life to the sub-microscopic domain, and forced us to use more abstract methods.

The first form of this new method was mainly based on spectroscopic evidence which led KRAMERS and HEISENBERG to the conviction that the proper description of the transition between two stationary states cannot be given in terms of the harmonic components of these states separately, but needs a new kind of transition quantity, depending on both states. HEISENBERG's quantum mechanics of 1926 is the first formulation of rules to handle these transition quantities, and these rules were soon recognized by myself as being identical with the matrix calculus of the mathematicians. This theory was developed by HEISENBERG, JORDAN and myself, and independently, in a most general and perfect manner, by DIRAC.

Again, independently SCHRÖDINGER developed in 1926 DE BROGLIE's wave mechanics by establishing a wave equation valid not only for free electrons but also for the case of external fields and mutual interaction, and showed its complete equivalence with matrix mechanics.

Concerning the physical interpretation, SCHRÖDINGER thought one ought to abandon the particle conception of the electron completely and to replace it by the assumption of a vibrating continuous cloud. When I suggested that the square of the wave function should be interpreted as probability density of particles, and produced evidence for it by a wave theory of collisions and other arguments, I found not only SCHRÖDINGER in opposition, but also, strangely enough, HEISENBERG. On the other hand, DIRAC developed the same idea in a mathematically brilliant way, which was soon generally accepted, also by HEISENBERG who produced a most important contribution by formulating his uncertainty relations (1927). These paved the way for a deeper philosophical analysis of the foundations of the new theory, achieved by BOHR's principle of complementarity, which replaces to some degree the classical concept of causality.

In a very short time the new theory was well established by its successes. I can mention only a few points: PAULI's matrix representation of the spin and DIRAC's relativistic generalization (1928) which led to the prediction of the positron, actually found by ANDERSON in 1932. Then came the systematic theory of the electronic structures of atoms and molecules and their relations to line and band spectra, to magnetism and other phenomena. WIGNER showed in 1927 how the general features of atomic structures could be found with the help of group theory. HARTREE, FOCK, HYLLERAAS and others developed numerical methods. The theory of collisions of atoms with electrons and other atoms was started by myself and developed by BETHE, MOTT, MASSEY and others, from which finally sprang a general theory of the penetration of particles through matter by BOHR.

Further, the derivation of the nature of the chemical bond, initiated by HEITLER and LONDON in 1927, was worked out by HUND, SLATER, MULLIKEN, PAULING and others. Even the complicated phenomena of reaction velocities, including catalytic acceleration, have been reduced to quantum mechanics.

Finally came DIRAC's most important theory of emission, absorption and scattering of electromagnetic radiation which led to the first systematic attempt of formulating quantum electrodynamics by FERMI, JORDAN, HEISENBERG and PAULI (1929), and later to the general theory of quantized fields and their interaction (WENTZEL, ROSENFELD, from 1931).

The last period of our fifty years is dominated by nuclear physics. Although the importance of nuclear research is probably greater than that of any other branch of physics, I shall be rather short about it for it is the most recent phase of our science and scarcely yet history.

The first breaking up of a nucleus was achieved by RUTHERFORD in 1919, by bombarding nitrogen with α-rays. Artificially accelerated particles were first used by COCKCROFT and WALTON in 1930. At that time the nucleus was believed to be composed of protons and electrons. But this led to difficulties if one tried to derive the angular momenta of nuclei from the spins of the component particles. In 1932 CHADWICK discovered the neutron, and those difficulties disappeared if the nucleus was assumed to consist of protons and neutrons, or charged and uncharged 'nucleons.' FERMI showed in 1932 that neutrons are most efficient in disrupting nuclei as they are not repelled by the nuclear charge. Many of the residual nuclei were found by IRENE and FREDERIC JOLIOT-CURIE in 1934 to be radioactive themselves.

The continuous β-ray spectrum offered great difficulties to the understanding until PAULI, in 1931, suggested the existence of the neutrino and FERMI developed, in 1934, the neutrino theory of the β-decay where the laws of conservation of energy and momentum are preserved. The line spectrum of β-rays was recognized to be of secondary origin, namely, due to the expulsion of electrons from the electronic cloud by γ-rays emitted by the nucleus.

The need for fast projectiles was first supplied by the use of cosmic rays. These had been discovered by HESS already in 1912 and their study has grown today into a vast science, covering not only nuclear physics but also geophysics, astronomy and cosmology.

The artificial production of fast particles has made enormous strides through the construction of powerful accelerating machines, as that of VAN DE GRAAFF (1931), LAWRENCE's cyclotron (1931), KERST's betatron (1940) and combinations of these, like the synchrotron.

The clue to the interpretation of nuclear transformations is EINSTEIN's formula $E = mc^2$, or more precisely, the relativistic conservation laws of energy and momentum. I am not an expert in the new awe-inspiring science of nuclear chemistry and shall make no attempt to describe it. I can say only a few words on the theoretical problems of nuclear physics. It is remarkable how many important facts can be understood by extremely simple models, as, for example, GAMOW's crater model (1928), which explains the α-decay and the GEIGER-NUTTAL relation between α-energy and life-time; and the liquid-drop model, suggested by VON WEIZSÄCKER in 1925, to explain the mass defect (nuclear energy) curve and later used successfully by BOHR (1935) to explain the mechanism of capture, re-emission and fission. A great amount of work has been done on exact quantum-mechanical calculations of the structure and properties of light nuclei (in particular the deuteron) and of the effect of collisions, with the aim of learning something about the nuclear forces. Important results have been obtained, but altogether the situation is not satisfactory.

Quite independently of detailed theories, the empirical values of the nuclear masses (internal energies) indicate that the light nuclei have the tendency to fuse, the heavy ones to disintegrate; hence all matter, except the elements in the middle of the periodic system (iron), is in principle unstable. But reaction velocities are, under terrestrial conditions, so extremely slow that nothing happens. It is, however, different in the interior of stars; BETHE showed in 1938 that one can account for the heat developed in the sun and the stars by a nuclear catalytic chain reaction, the fusion of four nucleons to form a helium nucleus.

The opposite phenomenon, the fission of the heavy nucleus of uranium into almost equal parts, discovered by HAHN and STRASSMANN in 1938, has initiated a new era in the sociological situation of our science and very likely in the history of mankind. Here is a list of events:

The establishment in 1939 of the possibility of a self-supporting chain reaction by different authors (JOLIOT, HALBAN and KOVARSKI; FERMI; SZILLARD); the construction of the first nuclear reactor or 'pile' under the direction of FERMI in 1942, and, finally, the harnessing of the industrial power of the United States to produce the atom bomb.

The political and economic implications of this development are too formidable to be discussed here; but I cannot refrain from saying that I, personally, am glad not to have been involved in the pursuit of research which has already been used for the most terrible mass destruction in history and threatens humanity with even worse disaster. I think that the applications of nuclear physics to peaceful ends are a poor compensation for these perils.

However, the human mind is adaptable to almost any situation. So let us forget for a while the real issues and enjoy the useful results obtained from the pile. In physics the remaining few gaps of the Periodic Table have been filled and five or six transuranium elements (among them fissionable nuclei like neptunium and plutonium) discovered. Innumerable new isotopes of known elements have been produced. Some of these can be used as 'tracers' in chemical and biological research as first suggested by VON HEVESY in 1913; others as substitute for the expensive radium in industrial research and in the treatment of cancer.

From the point of view of natural philosophy the most important achievement of the past decade seems to me the discovery of the meson, theoretically predicted by YUKAWA in 1935, which showed how far we are still removed from a knowledge of the real fundamental laws of physics. YUKAWA became convinced that the forces between nucleons are at least as important as the electromagnetic forces, and by applying the field concepts in analogy to MAXWELL's theory was able to predict a new particle which has the same relation to the nuclear field as the photon to the electromagnetic field, but has a finite rest-mass, which from the range of nuclear forces could be estimated to be about 300 electron masses. Soon the existence of mesons was experimentally confirmed in cosmic rays by ANDERSON and NEDDER-MEYER in 1936 and later with particles produced by the cyclotron in California in 1948. The method of photographic tracks has, in the hands of POWELL (from 1940) and others, produced a wealth of new results, for example, the spontaneous disintegration of the meson of about 300 electron masses into a lighter one of about 200 electron masses and a neutral particle. A meson of about 900 electron masses has been fairly well established, and it is not unlikely that still more types exist.

It is obvious that to understand all this a much deeper research in the theory of quantized fields and their interaction is necessary. A revised and modernized quantum electrodynamics was published independently by SCHWINGER in the United States and TOMONAGA in Japan in 1947, and from this has sprung a considerable amount of literature, aiming at the elimination of divergence difficulties and calculating effects of higher order, inaccessible to the older theory. A great success was the explanation of an observation made by LAMB and RUTHERFORD in 1947, which showed that DIRAC's celebrated theory of the hydrogen spectrum is not quite correct. But it becomes more and more clear that all these mathematical refinements do not suffice, and that a far more general theory has to be found, in which a new constant (an absolute length—or time, or mass) appears and which ought to account for the masses found in Nature. I wish to end this outlook into the future with a remark I have recently heard from HEISENBERG. We have accustomed ourselves to abandon deterministic causality for atomic events; but we have still retained the belief that probability spreads in space (multidimensional) and time according to deterministic laws in the form of differential equations. Even this has to be given up in the high-energy region. For it is obvious that the absolute time interval restricts the possibility of distinguishing the time order of events. If this interval is defined in the rest system it becomes large in a fast-moving system according to the relativistic time expansion (in contrast to the contraction of length).

Hence the indeterminacy of time order and therefore of cause-effect relation becomes large for fast particles.

Thus experience again leads us to an alteration of the metaphysical foundations of a rather unexpected kind. In fact, traditional philosophy has provided the leaders of our science, like EINSTEIN, BOHR and HEISENBERG, with problems in so far as it failed to supply answers agreeing with experience. I am convinced that although physics free from metaphysical hypotheses is impossible, these assumptions have to be distilled out of physics itself and continuously adapted to the actual empirical situation. On the other hand, the continuity of our science has not been affected by all these turbulent happenings, as the older theories have always been included as limiting cases in the new ones. The scientific attitude and the methods of experimental and theoretical research have been the same all through the centuries since GALILEO and will remain so.

IS CLASSICAL
MECHANICS IN FACT
DETERMINISTIC?

[First published in *Physikalische Blätter*, Vol. ii, No. 9, 49–54 (1955).]

The laws of classical mechanics, and through them the laws of classical physics as a whole, are so constructed that, if the variables in a closed system are given at some initial point of time, they can be calculated for any other instant—in principle, at least; for it is in most cases beyond human ability to carry out the mathematics involved. This deterministic idea has greatly attracted many thinkers, and has become an essential part of scientific philosophy. Modern physics, however, has been compelled to abandon determinism, together with other time-honored theories of space, time and matter, under the pressure of new empirical discoveries. Quantum mechanics, which has taken over the place of Newtonian mechanics, allows only statistical statements concerning the behaviour of mass particles. The great majority of physicists have become reconciled to this state of affairs, for it corresponds exactly to the empirical situation in atomic and nuclear physics, where experiments are based fundamentally on the counting of events. Among the theoreticians, however, there are some who are not content, and they are indeed some of the great ones to whom the quantum theory owes its origin and development. So far as I know, PLANCK himself was always sceptical towards the statistical interpretation of quantum mechanics. The same is true of EINSTEIN; even today he continues to point out, by means of ingenious examples, contradictions in this interpretation (and he is, moreover, still more concerned with the resolution of the concept of physical reality, which is closely involved with the problem of determinism). SCHRÖ-DINGER goes still further; he proposes to abandon the concept of particles (electrons, nuclei, atoms, etc.) and to construct the whole of physics upon the idea of waves, which obey deterministic laws in accordance with wave mechanics. DE BROGLIE (and others) take the opposite course; they reject waves, and seek a reinterpretation of quantum mechanics, in which everything is in principle determinate, and an uncertainty in prediction arises only by the presence of concealed and unobservable parameters. None of these physicists denies that quantum mechanics within the realm of its validity, i.e., apart from the theory of elementary particles, is in agreement with experiment and meets all the demands of the experimenters. Their

78

rejection is in every case founded on the assertion that the usual interpretation of the quantum formulae is obscure and philosophically unsatisfactory.

What now is this philosophy? I do not think it can be traced back before GALILEO and NEWTON. There were, of course, predictions before that in astronomy, of conjunctions and eclipses, but the men of antiquity and the Middle Ages saw order and predetermination only in the celestial spheres, whilst caprice and chaos reigned on earth. The religious tenets of fate and predestination relate not to the processes of Nature, but to Man, and are certainly fundamentally different from the mechanical determinism which we here consider. The latter is inconceivable without NEWTON's laws of motion and their astonishing success in the prediction of celestial events; it was derived from these laws, and later, during the eighteenth and nineteenth centuries, became a fundamental creed in science as a whole. The remarkable thing here is that the undoubted fact that Newtonian mechanics does not suffice to account for the observations, particularly in atomic physics, is inadequate to shake belief in this abstract theorem.

But is it certain that classical mechanics in fact permits prediction in all circumstances?* My doubts of this increase when I compare the time scales of astronomy and atomic physics. The age of the universe is reckoned to be some 10^9 years, i.e., orbital periods of the Earth. The number of periods in the ground state of the hydrogen atom, on the other hand, is of the order of 10^{16} per second. Thus, when time is measured in the units appropriate for each case, the situation is exactly the opposite of the simple conception: the stellar universe is short-lived, and the atomic universe extremely long-lived. Is it not dangerous to draw, from experience of the short-lived universe, conclusions which are to be valid for the long-lived one also?

These doubts are intensified when one considers the kinetic theory of gases. It is usually asserted in this theory that the result is in principle determinate, and that the introduction of statistical considerations is necessitated only by our ignorance of the exact initial state of *a large number* of molecules. I have long thought the first part of this assertion to be extremely suspect. Let us consider the simple case of a moving spherical molecule, which rebounds elastically from numerous other fixed molecules (a kind of three-dimensional bagatelle). A very small change in the direction of the initial velocity will then result in large changes of the path in the zigzag motion; for a small angular change brings about larger and larger spatial deviations, and so it must finally happen that a sphere which was formerly hit is now missed. If the initial deviation in direction is reduced, the moment when the path is changed to another is delayed, but it will occur eventually. If we require

* The question was raised already by R. v. MISES; s. p. 17, article "On the Meaning of Physical Theories," p. 34.

determinacy for all times, the smallest deviation in the initial direction must be avoided.* But has this any physical meaning? I am convinced that it has not, and that systems of this kind are in fact indeterminate. To justify this assertion, a clear comprehension of the idea of determination is needed.

First of all, we may distinguish between dynamical stability and instability. A motion is said to be stable if a small change Δx_0, Δv_0 in the initial state (where x denotes the set of all co-ordinates and v that of all velocities) causes only a small change Δx, Δv in the final state (so that, for all times, $\Delta x < M \Delta x_0$, $\Delta v < M \Delta v_0$, where M is a constant of the order of unity). Otherwise the motion is said to be unstable. It is fairly certain that the motion of the spheres in the bagatelle game discussed above is unstable. (This will be true *a fortiori* for a gas consisting of many moving elastic particles.) The question has been much argued as to whether or not the motion of the planets is stable. I do not know what is the result of modern research (theory of the three-body and many-body problems); it is of no importance for our purposes. The essential thing is that there are systems which serve as models of physical processes, and which, firstly, remain within a finite region of space and for which, secondly, all motions are dynamically unstable. The gas model which consists of elastic spheres in a container with elastic walls is probably such a system, but it is too complex to be analyzed rigorously. It is sufficient to consider the following trivially simple case. A mass particle moves without friction along a straight line (the x-axis) under no forces, and is elastically reflected at the termini ($x = 0$, $x = l$). The co-ordinate x remains in the finite interval $0 < x < l$ for any initial state (x_0, v_0), the velocity v remains constant, but the deviation Δx increases with time ($\Delta x = \Delta x_0 + t \Delta v_0$) and takes arbitrarily large values at sufficiently remote times. Thus any motion is unstable.

The connection with the problem of determinism is now evident. If we wish to retain the assertion that in this system the initial state determines every other state, we are compelled to demand absolutely exact values of x_0, v_0, and to prohibit any deviation Δx_0, Δv_0. We could then speak of 'weak' determinacy, as opposed to the 'strong' case where all motions are dynamically stable, and therefore predictions are actually possible. This, however, would be a mere evasion. The true situation is this. After a critical time $t_c = l/\Delta v_0$ has been reached, the uncertainty $\Delta x > l$, and the mass point may be found anywhere in the interval $0 < x < l$. That is to say, the final position is undetermined. If, however, Δv_0 is reduced, the critical time t_c is only delayed; it remains finite for any finite Δv_0, and becomes infinite only for $\Delta v_0 = 0$, i.e., for an absolutely definite initial velocity.

The connection with the problem of the continuum is evident here. An exhaustive

---

* We are evidently dealing with a double limit: the number of collisions tends to infinity, while the change in direction tends to zero; the result is undetermined in the absence of further data.

discussion of this question would take us too far afield, and the following brief must suffice. Statements like 'a quantity x has a completely definite value' (expressed by a real number and represented by a point in the mathematical continuum) seem to me to have no physical meaning. Modern physics has achieved its greatest successes by applying a principle of methodology, that concepts whose application requires distinctions that cannot in principle be observed, are meaningless and must be eliminated. The most striking examples are EINSTEIN's foundation of the special and general theories of relativity (of which the first rejects the concept of absolute simultaneity, and the second the distinction between gravity and acceleration as unobservable), and HEISENBERG's foundation of quantum mechanics (by eliminating the unobservable orbital radii and frequencies from BOHR's theory of the atom). The problem of continuity calls for the application of the same principle. A statement like $x = \pi$ cm. would have a physical meaning only if one could distinguish between it and $x = \pi_n$ cm. for every n, where π_n is the approximation of π by the first n decimals. This, however, is impossible; and even if we suppose that the accuracy of measurement will be increased in the future, n can always be chosen so large that no experimental distinction is possible.

Of course, I do not intend to banish from physics the idea of a real number. It is indispensable for the application of analysis. What I mean is that a physical situation must be described by means of real numbers in such a way that the natural uncertainty in all observations is taken into account.

Fifty years ago, FELIX KLEIN called for a similar step to be taken in geometry. Besides abstract, exact geometry, he desired to have a practical geometry, in which a point is replaced by a small spot, straight lines by narrow strips, etc. However, nothing much resulted from this. In the meantime, physics has independently developed the necessary tool, namely physical statistics. The statement 'x is equal to a real number' is replaced by 'the probability that x lies in an interval $x_1 < x < x_2$ is $P(x_1 | x | x_2)$.' Here x, x_1, x_2, P can be regarded as real numbers, since this is analytically convenient, whilst the exact measurability of quantities is not involved; P represents only the approximate expectation when cases are counted for which x is limited approximately by x_1 and x_2. In other words, the true physical variable is the probability density $P(x)$.

Quantum mechanics has realized that this is the only possible description of physical situations. (However, by introducing probability amplitudes, it goes far beyond this statistical viewpoint.)

In classical mechanics, the statistical method is used only for systems of very many individual particles. Our model shows that it is obligatory to use it in every case, even that of a single particle in the simplest conceivable conditions. This does not require any new mathematical considerations; for the law whereby the probability density varies is given at once by LIOUVILLE's theorem in mechanics.* I shall elsewhere discuss exhaustively the mathematical details and the relation to quantum mechanics. Here I shall briefly give some results.

If we first continue to use classical mechanics, we find that our model is perhaps the simplest example of the so-called ergodic theorem of statistical mechanics. It can be very easily shown that an initial probability density, describing an almost definite state, passes in time into what is called the microcanonical distribution. This therefore occurs automatically, even for *one particle,* and has nothing to do with the 'large number' of particles. Complex systems with energy exchange need be taken into account only if we wish to pass to the canonical distribution.

Now, the same model can also be treated by quantum mechanics. An initial state with an uncertainty Δx_0 in the initial position is then described by a wave packet; the uncertainty Δv_0 in the initial velocity cannot be supposed arbitrarily small, but is related to Δx_0 by HEISENBERG's uncertainty relation $\Delta x_0 . \Delta v_0 > \hbar/2m$; this holds for all times, the factors Δx and Δv varying with time. If both Δx_0 and Δv_0 can be made small (for large masses), the quantum formulae are identical with the classical ones to a close approximation, and there is again a critical instant t_c where the individual motion ceases and a state is entered which can be described only statistically. This corresponds exactly to the usual description of a motion, in quantum mechanics, by means of stationary waves, which is thus the analogue of the classical microcanonical distribution.

To summarize, we may say that it is not the introduction of the indeterministic statistical description which places quantum mechanics apart from classical mechanics, but other features, above all the concept of the probability density as the square of a probability amplitude $P = |\Psi|^2$; the phenomenon of probability interference results from this, and therefore it is impossible to apply without modification the idea of an 'object' to the mass particles of physics: the concept of physical reality must be revised. This, however, is beyond the scope of these elementary considerations.

APPENDIX

LIOUVILLE's theorem expresses the conservation of probability density during the motion, and leads to the differential equation

$$\frac{\partial P}{\partial t} = \frac{\partial H}{\partial x}\frac{\partial P}{\partial p} - \frac{\partial H}{\partial p}\frac{\partial P}{\partial x}, \qquad . \qquad . \qquad . \qquad (1)$$

where H is HAMILTON's function. (The expression on the right is the so-called

---

* See Appendix. Also *Proceedings of the Danish Academy,* 30, No. 2, 1955. (Festskript til Niels Bohr.)

Poisson bracket.) The solution corresponding to an initial state $P(x, p, 0) = F(x, p)$ is

$$P(x, p, t) = F[f(x, p, t), g(x, p, t)], \quad . \quad . \quad . \quad (2)$$

where $f(x, p, t) = $constant, $g(x, p, t) = $constant are two integrals of the canonical equations of motion, normalized so that

$$f(x, p, 0) = x, p, 0) = p. \quad . \quad . \quad . \quad (3)$$

The solution of the probability equation (1) and of the canonical equations thus present entirely equivalent problems. Nevertheless, the solution of (1) furnishes new and interesting results.

For the example given in the text we have $H = p^2/2m$; thus (1) becomes

$$\frac{\partial P}{\partial t} = v \frac{\partial P}{\partial x}, \quad (v = p/m). \quad . \quad . \quad . \quad (4)$$

Two normalized integrals are $f = x - vt$, $g = v$, and so the solution (2) is

$$P = F(x - vt, v). \quad . \quad . \quad . \quad (5)$$

The boundary conditions amount to the requirement of periodicity in x (with period $2l$) and antisymmetry in x and v:

$$F(x + 2l, v) = F(x, v), \quad F(-x, -v) = F(x, v) \quad . \quad . \quad (6)$$

This can be satisfied with an arbitrary function $f(x, v)$ by

$$F(x, v) = \sum_{k=-\infty}^{\infty} [f(2kl + x, v) + f(2kl - x, -v)]. \quad . \quad . \quad (7)$$

If we here replace x by $x - vt$ according to (5), we obtain $P(x, v, t)$. If the position and velocity at the initial instant are almost definite, $f(x, v)$ must be taken as a function having a sharp maximum at (x_0, v_0) and vanishingly small elsewhere. If f is a Gaussian function in both x (width σ_0) and v (width τ_0), the resultant x-distribution

$$P(x, t) = \int P(x, v, t) dv \quad . \quad . \quad . \quad (8)$$

is again a sum of Gaussian functions in x with width

$$\sigma(t) = \sqrt{(\sigma_0^2 + \tau_0^2 t^2)} \quad . \quad . \quad . \quad (9)$$

which varies as t when t is large.

This passage to the limit $t \to \infty$ can be simply described by drawing a small circle round the point (x_0, v_0) in the (x, p) phase space (or the xv-plane), and examining how this breaks up into two ellipses of equal area with centres $x_0 \pm v_0 t$, whose major axes become more and more parallel to the x-axis and finally longer than the interval l.

ASTRONOMICAL
RECOLLECTIONS

[First published in *Vistas in Astronomy*, Vol. I, pp. 41–44 (1955), Pergamon Press, London. This work is dedicated to Professor F. J. M. STRATTON for the occasion of his seventieth birthday.]

I am not an astronomer, nor have I done any work in physics applicable to astronomy. Yet I cannot resist the wish to be included amongst those who offer their congratulations to Professor STRATTON by an article in this volume. There was a time in my life when I was very near to devoting myself to the celestial science; but I failed. May I offer, as a substitute for a more serious contribution, the story of my wrangling with astronomy and some recollections of remarkable astronomers who were my teachers.

I have to begin with Professor FRANZ, the director of the observatory of my home city, Breslau. My father, who died just before I finished school, had left me the advice to attend lectures on various subjects before choosing a definite study for a profession. In Germany at that period this was possible because of the complete 'academic freedom' at the university.

There was in most subjects no strict syllabus, no supervision of attendance, no examinations except the final ones. Every student could select the lectures he liked best; it was his own responsibility to build up a body of knowledge sufficient for the final examinations which were either for a professional certificate or for a doctor's degree, or both. Thus I made up a rather mixed programme for my first year, including physics, chemistry, zoology, general philosophy and logic, mathematics and astronomy. At school I had never been very good nor interested in mathematics, but at the university the only lectures which I really enjoyed were the mathematical and astronomical ones. The greatest disappointment were the philosophical courses; there we heard a lot about the rules of rational thinking, the paradoxes of space, time, substance, cause, the structure of the universe, and infinity. Yet it seemed to me an awful muddle. Now the same concepts appeared also in the mathematical and astronomical lectures, but instead of being veiled in a mist of paradox they were formulated in a clear way according to the case. For that was the important discovery I then made: that all the high-sounding words connected with the concept of infinity mean nothing unless applied in a definite system of ideas to a definite problem.

Astronomy was attractive in another way. There the problems of cosmology are related to the infinity of the physical universe. But little about these great questions was mentioned in the elementary lectures of our Professor FRANZ. What we had to learn was the careful handling of instruments, correct reading of scales, elimination of errors of observation and precise numerical calculations—all the paraphernalia of the measuring scientist. It was a rigorous school of precision, and I enjoyed it. It gave one the feeling of standing on solid ground. Yet actually this feeling was not quite justified by facts. The Breslau observatory was not on solid ground, but on the top of the high and steep roof of the lovely university building, in a kind of roof pavilion, decorated with fantastic baroque ornaments and statues of saints and angels. The main instrument was a meridian circle, which a hundred years ago had been used by the great BESSEL; although it was placed on a solid pillar standing on the foundations and rising straight through the whole building, it was not free from vibrations produced by the gales blowing from the Polish steppes. The whole outfit of this observatory was old-fashioned and more romantic than efficient. There were several old telescopes from WALLENSTEIN's time, like those KEPLER may have used. We had no electric chronograph but had to learn to observe the stars crossing the threads in the field of vision by counting the beats of a big clock and estimating the tenths of a second. It was a very good school of observation, and it had the additional attraction of an old and romantic craft.

I remember many an icy winter's night spent there in the little roof pavilion. We were only three students in astronomy, and we took the observations alternately. When my turn was finished I enjoyed looking down on the endless expanse of snow-covered, gabled roofs of the ancient city, the silhouettes against the starry sky of the Cathedral further away beyond the river. There on the narrow balcony amongst the stucco saints and old-fashioned telescopes, one felt like an adept of Dr. Faustus and would not have wondered if Mephistopheles had appeared behind the next pillar. However, it was only old Professor FRANZ who came up the steps to look after his three students—he had not had so many for a long time—and who carried with him the soberness of the exact scientist, checking our results and criticizing our endeavours with mild and friendly irony.

These, our results, I rather think were not very reliable; it was not so much our fault as that of the exalted but exposed position of the observatory. Professor FRANZ himself, therefore, abstained from doing research, which needed exact measurements, and restricted himself to descriptive work, a thorough study of the moon's surface which he knew better than the geography of our own planet. He made strenuous efforts, however, to obtain a modern observatory but never succeeded. During my student time there were great hopes. The firm Carl Zeiss, Jena, had sent a set of modern instruments to the World's Fair at Chicago. After the end of the show these were purchased by the Prussian State for its university observatories. Breslau obtained an excellent meridian instrument and a big parallactic tele-scope; yet no proper building was granted, and the meridian circle was installed in a wooden cabin on a narrow island of the Oder River, just opposite the university building. This island was in fact an artificial dam between the river and a lock through which many barges used to pass. The time service for the province of Silesia, which had been practised for scores of years with the help of the old BESSEL circle, was transferred to the new Zeiss instruments, but the results remained highly unsatisfactory. Eventually we discovered a correlation between the strange ir-regularities of the time observations with the changing level of the water in the lock; the island suffered small displacements through the water pressure. Professor FRANZ's hopes of a more efficient observatory had broken down again.

We youngsters took this disappointment rather as a funny incident. It did not diminish the fascination which astronomy exerted on my mind. This fascination was, however, shattered by the horrors of computation. FRANZ gave us a lecture on the determination of planetary orbits, connected with a practical course where we had to learn the technique of computing, filling in endless columns of seven decimal logarithms of trigonometric functions according to traditional forms. I knew from school that I was bad at numerical work, but I tried hard to improve. It was in vain, there was always a mistake somewhere in my figures, and my results differed from those of the class mates. I was teased by them, but that made it worse. I do not think that I ever finished an orbit or an ephemeris, and then I gave up—not only this calculating business but the whole idea of becoming an astronomer. If I had known at that time that there was in existence another kind of astronomy which did not consider the prediction of planetary positions as the ultimate aim, but studied the physical structure of the universe with all the powerful instruments and con-cepts of modern physics, my decision might have been different. But I came in con-tact with astrophysics only some years later, when it was too late to change my plans.

At that period German students used to move from one university to another, from different motives. Sometimes they were attracted by a celebrated professor or a well-equipped laboratory, in other cases by the amenities and beauties of a city, by its museums, concerts, theatres, or by winter sport, by carnival and gay life in general. Thus I spent two summer semesters in Heidelberg and Zürich, returning during the winter to the home university. The observatory of Heidelberg was on the Königstuhl, a considerable, wooded hill, where the astronomers lived a secluded life remote from the ordinary crowd. I had then definitely changed over to physics, and not even the celebrated name of WOLF, the professor who has discovered more planetoids than anybody else, deflected me from my purpose.

The observatory in Zürich was more accessible, and the name of the professor was Wolfer, which could be interpreted as a comparative to Wolf. But even that did not attract me.

The following summer I went to Göttingen for the rest of my student time. There Karl Schwarzschild was director of the famous observatory which had been for many years under the great Gauss. Schwarzschild was the youngest professor of the university, about thirty years of age; a small man with dark hair and a moustache, sparkling eyes and an unforgettable smile. I joined his astrophysical seminar and was for the first time introduced to the modern aspect of astronomy. We discussed the atmosphere of planets, and I had to give an account of the loss of gas through diffusion against gravity into interstellar space. Thus I was driven to a careful study of the kinetic theory of gases which then, in 1904, was not a regular part of the syllabus in physics. But this is not the only subject which I first learned through Schwarzschild's teaching. His was a versatile, all-embracing mind, and astronomy proper only one field of many in which he was interested. About this time he published deep investigations on electro-dynamics, in particular on the variational principle from which Lorentz' equations for the field of an electron and for its motion could be derived. In the following year (1905) there appeared the first of his great articles on the aberrations of optical instruments; these are, in my opinion, classical investigations, unsurpassed in clarity and rigour by later work. I have presented this method in my book *Optik* (Springer, 1932), and it is again the backbone of a modernized version which appeared as an English book on optics (in collaboration with E. Wolf*). Schwarzschild applied his aberration formulae to the actual construction of new types of optical systems; but I am not competent to speak about this part of his activities. Nor can I discuss his astronomical work, experimental or theoretical. Personally he was a most charming man, always cheerful, amusing, slightly sarcastic, but kind and helpful. He once saved me from an awkward situation. I had intended to take geometry as one of my subjects in the oral examinations for the doctor's degree, but was not attracted by the lectures of Felix Klein, the famous mathematician, and attended somewhat irregularly. This fact did not escape Klein's observation and he showed me his displeasure. A disaster at the orals, only six months ahead, seemed to be impending. But Schwarzschild said that half a year was ample time to learn the whole of astronomy. He gave me some books to read and tutored me a little, in exchange for my training him in tennis. When the examination came his first question was: 'What do you do when you see a falling star?' Whereupon I answered at once: 'I make a wish'—according to an old German superstition that such a wish is always fulfilled. He remained quite serious and continued: 'Yes, and what do you do then?' Whereupon I gave the

---

* *Principles of Optics*. Third revised edition. London: Pergamon Press, 1966.

expected answer: 'I would look at my watch, remember the time, constellation of appearance, direction of motion, range, etc., go home and work out a crude orbit.' Which led to celestial mechanics and to a satisfactory pass. SCHWARZSCHILD differed from the ordinary type of the dignified, bearded German scholar of that time; not only in appearance, but also in his mental structure, which was thoroughly modern, cheerful, active, open to all problems of the day. Still he had his hours of professorial absent-mindedness. There was a 'Stammtisch,' a certain table in a restaurant where a group of young professors and lecturers used to meet for lunch. SCHWARZSCHILD was one of them until his marriage. A few weeks after the wedding he was again at his accustomed place at the lunch table and plunged in his usual way into a lively discussion about some scientific problem, until one of the men asked him: 'Now, SCHWARZSCHILD, how do you like married life?' He blushed, jumped up, said, 'Married life—oh, I have quite forgotten—,' got his hat and ran away. But I think this kind of behavior was not typical of him. He always knew what he was doing. His life was short, his achievements amazing, his success great—his end tragic. When the great war of 1914–18 broke out he was employed as a mathematical expert in ballistics and attached to the staff of one of the armies on the Eastern front. There, in Russia, he contracted some rare infectious disease. It was said that he refused to be sent home, until it was too late. On his way home, he visited me in my military office in Berlin; he was still cheerful, but he looked terribly ill. Soon after he died. Now his son, Martin, keeps up the astronomical tradition, thus founding another one of those hereditary lines of astronomers, the HERSCHELS, the STRUVES, and so on.

I have met many other distinguished astronomers and been intimate with some of them; but as most of them are still wandering on this globe, I had better refrain from telling stories about them.

May I conclude by wishing Professor STRATTON many happy returns and by adding the request that he too may present us with some recollections of astronomical personalities out of his long experience.

STATISTICAL
INTERPRETATION OF
QUANTUM MECHANICS

[First published in *Science*, Vol. 122, No. 3172, pp. 675–679 (1955). This article is the English translation of the lecture Professor BORN gave in German when he was awarded the Nobel Prize for Physics in 1954, a prize which he shared with W. BOTHE.]

The published work for which the honor of the Nobel prize for the year 1954 has been accorded to me does not contain the discovery of a new phenomenon of nature but, rather, the foundations of a new way of thinking about the phenomena of nature. This way of thinking has permeated experimental and theoretical physics to such an extent that it seems scarcely possible to say anything more about it that has not often been said already. Yet there are some special aspects that I should like to discuss.

The first point is this: the work of the Göttingen school, of which I was at that time the director, during the years 1926 and 1927, contributed to the solution of an intellectual crisis into which our science had fallen through PLANCK's discovery of the quantum of action in the year 1900. Today physics is in a similar crisis—I do not refer to its implication in politics and economics consequent on the mastery of a new and terrible force of nature, but I am thinking of the logical and epistemological problems posed by nuclear physics. Perhaps it is a good thing to remind oneself at such a time of what happened earlier in a similar situation, especially since these events are not without a certain element of drama. In the second place, when I say that physicists had accepted the way of thinking developed by us at that time, I am not quite correct. There are a few most noteworthy exceptions—namely, among those very workers who have contributed most to the building up of quantum theory. PLANCK himself belonged to the sceptics until his death. EINSTEIN, DE BROGLIE, and SCHRÖDINGER have not ceased to emphasize the unsatisfactory features of quantum mechanics, and to demand a return to the concepts of classical, Newtonian physics, and to propose ways in which this could be done without contradicting experimental facts. One cannot leave such weighty views unheard. NIELS BOHR has gone to much trouble to refute the objections. I have myself pondered on them and believe I can contribute something to the clarification of the situation. We are concerned with the borderland between physics and philosophy, and so my physical lecture will be partly historically and partly philosophically colored, for which I ask indulgence.

First of all, let me relate how quantum mechanics and its statistical interpretation arose. At the beginning of the 1920's every physicist, I imagine, was convinced that PLANCK's hypothesis was correct, according to which the energy in oscillations of definite frequency ν (for example, in light waves) occurs in finite quanta of size $h\nu$. Innumerable experiments could be explained in this manner and always gave the same value of PLANCK's constant h. Furthermore, EINSTEIN's assertion that light quanta carry momentum $h\nu/c$ (where c is the velocity of light) was well supported by experiment. This meant a new lease of life for the corpuscular theory of light for a certain complex of phenomena. For other processes, the wave theory was appropriate. Physicists accustomed themselves to this duality and learned to handle it to a certain extent.

In 1913 NIELS BOHR had solved the riddle of line spectra by using quantum theory and at the same time had explained, in their main features, the wonderful stability of atoms, the structure of their electronic shells, and the periodic system of the elements. For the sequel the most important assumption of his teaching was this: an atomic system cannot exist in all mechanically possible states, which form a continuum, but in a series of discrete 'stationary' states; in a transition from one to another the difference in energy $E_m - E_n$ is emitted or absorbed as a light quantum $h\nu_{mn}$ (according as E_m is greater or less than E_n). This is an interpretation, in terms of energy, of the fundamental law of spectroscopy discovered some years previously by W. RITZ. The situation can be pictured by writing the energy levels of the stationary states twice over, horizontally and vertically; a rectangular array results

	E_1	E_2	E_3 $\dots$
E_1	11	12	13 $\dots$
E_2	21	22	23 $\dots$
$\dots$	$\dots$	$\dots$	$\dots$

in which positions on the diagonal correspond to the states and off-diagonal positions correspond to the transitions.

BOHR was fully aware that the law thus formulated is in conflict with mechanics and that, therefore, even the use of the concept of energy in this context is problematical. He based this bold fusion of the old with the new on his principle of correspondence. This consists in the obvious requirement that ordinary classical mechanics must hold to a high degree of approximation in the limit, when the numbers attached to the stationary states, the quantum numbers, are very large—that is, far to the right and low down in the foregoing array—so that the energy changes relatively little from place to place—that is, practically continuously.

Theoretical physics lived on this idea for the next 10 years. The problem was that a harmonic oscillator possesses not only frequency but intensity as well. For each transition in the scheme there must be a corresponding intensity. How is the latter to be found by considerations of correspondence? It was a question of guessing the unknown from a knowledge of a limiting case. Considerable success was achieved by BOHR himself, by KRAMERS, by SOMMERFELD, by EPSTEIN, and by many others. But the decisive step was again taken by EINSTEIN, who, by a new derivation of PLANCK's radiation formula, made it evident that the classical concept of intensity of emission must be replaced by the statistical idea of transition probability. To each position in our scheme there belongs, besides the frequency $\nu_{mn} = (E_m - E_n)/h$, a certain probability for the transition accompanied by emission or absorption of radiation.

In Göttingen we also took part in the attempts to distill the unknown mechanics of the atom out of the experimental results. The logical difficulty became ever more acute. Investigations on scattering and dispersion of light showed that EINSTEIN's conception of transition probability as a measure of the strength of an oscillation was not adequate, and the idea of an oscillation amplitude associated with each transition could not be dispensed with. In this connection work by LADENBURG [1], KRAMERS [2], HEISENBERG [3], JORDAN and I [4] may be mentioned. The art of guessing correct formulas, which depart from the classical formulas but pass over into them in the sense of the correspondence principle, was brought to considerable perfection. A paper of mine, which introduced in its title the expression 'quantum mechanics,' probably for the first time, contains a very involved formula—still valid at the present time—for the mutual disturbance of atomic systems.

This period was brought to a sudden end by HEISENBERG [5], who was my assistant at that time. He cut the Gordian knot by a philosophical principle and replaced guesswork by a mathematical rule. The principle asserts that concepts and pictures that do not correspond to physically observable facts should not be used in theoretical description. When EINSTEIN, in setting up his theory of relativity, eliminated the concepts of the absolute velocity of a body and of the absolute simultaneity of two events at different places, he was making use of the same principle. HEISENBERG banished the picture of electron orbits with definite radii and periods of rotation, because these quantities are not observable; he demanded that the theory should be built up by means of quadratic arrays of the kind suggested in a preceding paragraph. Instead of describing the motion by giving a co-ordinate as a function of time $x(t)$, one ought to determine an array of transition probabilities x_{mn}. To me the decisive part in his work is the requirement that one must find a rule whereby from a given array

$$x_{11} \qquad x_{12} \quad \cdot \quad \cdot \quad \cdot$$
$$x_{21} \qquad x_{22}$$
$$\cdot$$
$$\cdot$$
$$\cdot$$

the array for the square,

$$(x^2)_{11} \qquad (x^2)_{12} \quad \cdot \quad \cdot \quad \cdot$$
$$(x^2)_{21} \qquad (x^2)_{22}$$
$$\cdot$$
$$\cdot$$
$$\cdot$$

may be found (or, in general, the multiplication law of such arrays).

By consideration of known examples discovered by guesswork he found this rule and applied it with success to simple examples such as the harmonic and anharmonic oscillator. This was in the summer of 1925. HEISENBERG, suffering from a severe attack of hay fever, took leave of absence for a course of treatment at the seaside and handed over his paper to me for publication, if I thought I could do anything about it.

The significance of the idea was immediately clear to me, and I sent the manuscript to the *Zeitschrift für Physik*. HEISENBERG's rule of multiplication left me no peace, and after a week of intensive thought and trial, I suddenly remembered an algebraic theory that I had learned from my teacher, ROSANES, in Breslau. Such quadratic arrays are quite familiar to mathematicians and are called matrices, in association with a definite rule of multiplication. I applied this rule to HEISENBERG's quantum condition and found that it agreed for the diagonal elements. It was easy to guess what the remaining elements must be, namely, null; and immediately there stood before me the strange formula

$$pq - qp = h/2\pi i.$$

This meant that co-ordinates q and momenta p are not to be represented by the values of numbers but by symbols whose product depends on the order of multiplication—which do not 'commute,' as we say.

My excitement over this result was like that of the mariner who, after long voyaging, sees the desired land from afar, and my only regret was that HEISENBERG was not with me. I was convinced from the first that we had stumbled on the truth. Yet again a large part was only guesswork, in particular the vanishing of the nondiagonal elements in the foregoing expression. For this problem I secured the collaboration of my pupil PASCUAL JORDAN, and in a few days we succeeded in showing that I had guessed correctly. The joint paper by JORDAN and myself [6] contains the most important principles of quantum mechanics, including its extension to electrodynamics.

There followed a hectic period of collaboration among the three of us, rendered difficult by HEISENBERG's absence. There was a lively interchange of letters, my contribution to which unfortunately went amiss in the political disorders. The result was a three-man paper [7], which brought the formal side of the investigation to a certain degree of completeness. Before this paper appeared, the first dramatic surprise occurred: PAUL DIRAC's paper [8] on the same subject. The stimulus received through a lecture by HEISENBERG in Cambridge led him to results similar to ours in Göttingen, with the difference that he did not have recourse to the known matrix theory of the mathematicians but discovered for himself and elaborated the doctrine of such non-commuting symbols.

The first non-trivial and physically important application of quantum mechanics was made soon afterwards by W. PAULI [9], who calculated the stationary energy values of the hydrogen atom by the matrix method and found complete agreement with BOHR's formulas. From this moment there was no longer any doubt about the correctness of the theory.

What the real significance of this formalism might be was, however, by no means clear. Mathematics, as often happens, was wiser than interpretative thought. While we were still discussing the point, there occurred the second dramatic surprise: the appearance of SCHRÖDINGER's celebrated papers [10]. He followed quite a different line of thought, which derived from LOUIS DE BROGLIE [11]. The latter had a few years previously made the bold assertion, supported by brilliant theoretical considerations, that wave-corpuscle dualism, familiar to physicists in the case of light, must also be exhibited by electrons; to each freely movable electron there belongs, according to these ideas, a plane wave of perfectly definite wavelength, determined by PLANCK's constant and the mass. This exciting essay by DE BROGLIE was well known to us in Göttingen.

One day in 1925 I received a letter from C. J. DAVISSON containing singular results on the reflection of electrons from metallic surfaces. My colleague on the experimental side, JAMES FRANCK, and I at once conjectured that these curves of DAVISSON's were crystal-lattice spectra of DE BROGLIE's electron waves, and we arranged for one of our pupils, W. ELSASSER [12], to investigate the matter. His result provided the first quantitative proof of DE BROGLIE's idea, a proof independently given later by DAVISSON and GERMER [13] and by G. P. THOMSON [14], by systematic experiments.

But this familiarity with DE BROGLIE's line of thought did not lead on further toward an application to the electronic structure of atoms. This was reserved for SCHRÖDINGER. He extended DE BROGLIE's wave equation, which applied to free motion, to the case in which forces act and gave an exact formulation of the additional conditions, already hinted at by DE BROGLIE, to which the wave function Ψ must be subjected—namely, that it should be single-valued and finite in space and time—and he succeeded in deriving the stationary states of the hydrogen atom as monochromatic solutions of his wave equation not extending to infinity. For a short while, at the beginning of 1926, it looked as if suddenly there were two self-contained but entirely distinct systems of explanation in the field—matrix mechanics and wave mechanics. But SCHRÖDINGER himself soon demonstrated their complete equivalence.

Wave mechanics enjoyed much greater popularity than the Göttingen or Cambridge version of quantum mechanics. Wave mechanics operates with a wave function Ψ, which—at least in the case of one particle—can be pictured in space, and it employs the mathematical methods of partial differential equations familiar to every physicist. SCHRÖDINGER also believed that his wave theory made possible a return to deterministic classical physics; he proposed (and has emphatically renewed this suggestion quite recently, [15]) to abandon the particle picture entirely and to speak of electrons not as particles but as a continuous density distribution $|\Psi|^2$, or electric density $e|\Psi|^2$.

To us in Göttingen this interpretation appeared unacceptable in the face of the experimental facts. At that time it was already possible to count particles by means of scintillations or with the Geiger counter and to photograph their tracks with the help of the Wilson cloud chamber.

It appeared to me that it was not possible to arrive at a clear interpretation of the Ψ-function by considering bound electrons. I had therefore been at pains, as early as the end of 1925, to extend the matrix method, which obviously covered only oscillatory processes, in such a way as to be applicable to aperiodic processes. I was at that time the guest of the Massachusetts Institute of Technology in the U.S.A., and there I found in NORBERT WIENER a distinguished collaborator. In our joint paper [16] we replaced the matrix by the general concept of an operator and, in this way, made possible the description of aperiodic processes. Yet we missed the true approach, which was reserved for SCHRÖDINGER; and I immediately took up his method, since it promised to lead to an interpretation of the Ψ-function. Once more an idea of EINSTEIN's gave the lead. He had sought to make the duality of particles (light quanta or photons) and waves comprehensible by interpreting the square of the optical wave amplitudes as probability density for the occurrence of photons. This idea could at once be extended to the Ψ-function: $|\Psi|^2$ must represent the probability density for electrons (or other particles). To assert this was easy; but how was it to be proved?

For this purpose atomic scattering processes suggested themselves. A shower of electrons coming from an infinite distance, represented by an incident wave of known intensity (that is, $|\Psi|^2$) impinge on an obstacle, say a heavy atom. In the same way that the water wave caused by a steamer excites secondary circular waves in striking a pile, the incident electron wave is partly transformed by the atom into a secondary spherical wave, whose amplitude of oscillation Ψ is different in different directions. The square of the amplitude of this wave at a great distance from the scattering center then determines the relative probability of scattering in its dependence on direction. If, in addition, the scattering atom is itself capable of existing in different stationary states, one also obtains quite automatically from SCHRÖDINGER's wave equation the probabilities of excitation of these states, the electron being scattered with loss of energy, or inelastically, as it is termed. In this way it was possible to give the assumptions of BOHR's theory, first verified experimentally by FRANCK and HERTZ, a theoretical basis [17]. Soon WENTZEL [18] succeeded in deriving RUTHERFORD's celebrated formula for the scattering of α-particles from my theory.

But the factor that contributed more than these successes to the speedy acceptance of the statistical interpretation of the Ψ-function was a paper by HEISENBERG [19] that contained his celebrated uncertainty relationship, through which the revolutionary character of the new conception was first made clear. It appeared that it was necessary to abandon not only classical physics but also the naïve conception of reality that thought of the particles of atomic physics as if they were exceedingly small grains of sand. A grain of sand has at each instant a definite position and velocity. For an electron this is not the case; if one determines the position with increasing accuracy, the possibility of determining the velocity becomes less, and vice versa. I shall return to these questions in a more general connection, but before doing so would like to say a few words about the theory of collisions.

The mathematical techniques of approximation I used were somewhat primitive and were soon improved. Out of the literature, which has grown to unmanageable proportions, I can name only a few of the earliest authors, to whom the theory is indebted for considerable progress: HOLTSMARK in Norway, FAXÉN in Sweden, BETHE in Germany, MOTT and MASSEY in Great Britain.

Today collision theory is a special science, with its own voluminous textbooks, and has grown completely over my head. Of course, in the last resort all the modern branches of physics, quantum electrodynamics, the theory of mesons, nuclei, cosmic rays, elementary particles and their transformations, all belong to this range of ideas, to a discussion of which no bounds could be set.

I should also like to state that during the years 1926 and 1927 I tried another way of justifying the statistical conception of quantum mechanics, partly in collaboration with the Russian physicist FOCK [20]. In the aforementioned three-man paper there is a chapter in which the SCHRÖDINGER function is really anticipated; only it is not thought of as a function Ψ of space, but as function Ψ_n of the discrete index $n = 1, 2, \ldots$ which enumerates the stationary states. If the system under consideration is subject to a force that is variable in time, Ψ_n also becomes time-dependent, and $|\Psi_n(t)|^2$ denotes the probability for the existence of that state n at time t.

Starting from an initial distribution in which only one state is present, we obtain in this manner transition probabilities, and we can investigate their properties. In particular, what interested me most at the time was what happens in the adiabatic limiting case, that is, in the case of very slowly variable external action; it was possible to show that, as might have been expected, the probability of transitions became ever smaller. The theory of transition probabilities was developed independently by DIRAC and made to yield results. It may be said that the whole of atomic and nuclear physics works with this system of concepts, especially in the extremely elegant form given to them by DIRAC [21]; almost all experiments lead to statements about relative probabilities of events, even if they appear concealed under the name cross section or the like.

How then does it come about that great discoverers such as EINSTEIN, SCHRÖD-INGER, and DE BROGLIE are not satisfied with the situation? As a matter of fact, all these objections are directed not against the correctness of the formulas but against their interpretation. Two closely interwoven points of view must be distinguished: the question of determinism and the question of reality.

Newtonian mechanics is deterministic in the following sense. If the initial state (positions and velocities of all particles) of a system is accurately given, the state at any other time (earlier or later) may be calculated from the laws of mechanics. All the other branches of classical physics have been built up in accordance with this pattern. Mechanical determinism gradually became an article of faith—the universe as a machine, an automaton. As far as I can see, this idea has no precursors in ancient or mediaeval philosophy; it is a product of the immense success of Newtonian mechanics, especially in astronomy. In the nineteenth century it became a fundamental philosophic principle for the whole of exact science. I asked myself whether this was really justified. Can we really make absolute predictions for all time on the basis of the classical equations of motion? It is easily seen, by simple examples, that this is the case only if we assume the possibility of absolutely accurate measurement (of the position, velocity, or other quantities). Let us consider a particle moving without friction on a straight line between two end-points (walls) at which it suffers perfectly elastic recoil. The particle moves backward and forward with constant speed equal to its initial speed v_0, and one can say exactly where it will be at a stated time provided that v_0 is accurately known.

But if we allow a small inaccuracy Δv_0, the inaccuracy of the prediction of position at time t is $t\Delta v_0$; that is, it increases with t. If we wait long enough, until time

$t_c = l/\Delta v_0$, where c is the distance between the elastic walls, the inaccuracy Δx will have become equal to the whole interval l. Thus it is possible to say absolutely nothing about the position at a time later than t_c. Determinism becomes complete indeterminism if one admits even the smallest inaccuracy in the velocity datum. Is there any sense—I mean physical, not metaphysical, sense—in which one can speak of absolute data? Is it justifiable to say that the co-ordinate x is π cm, where $\pi = 3.1415 \ldots$ is the familiar transcendental number that determines the ratio of the circumference of a circle to its diameter? As an instrument of mathematics, the concept of a real number represented by a nonterminating decimal is extremely important and fruitful. As a measure of a physical quantity, the concept is nonsensical. If the decimal for π is interrupted at the 20th or 25th place, two numbers are obtained which cannot be distinguished by any measurement from each other and from the true value. According to the heuristic principle employed by EINSTEIN in the theory of relativity and by HEISENBERG in quantum theory, concepts that correspond to no conceivable observation ought to be eliminated from physics. This is possible without difficulty in the present case also; we have only to replace statements like $x = \pi$ cm. by: the probability of the distribution of values of x has a sharp maximum at $x = \pi$ cm.; and (if we wish to be more accurate) we can add: of such and such a breadth. In short, ordinary mechanics must be formulated statistically. I have occupied myself with this formulation a little recently and have seen that it is possible without difficulty. This is not the place to go into the matter more closely. I only wish to emphasize the point that the determinism of classical physics turns out to be a false appearance, produced by ascribing too much weight to mathematicological conceptual structures. It is an *idol*, not an *ideal*, in the investigation of nature and, therefore, cannot be used as an objection to the essentially indeterministic, statistical interpretation of quantum mechanics.

Much more difficult is the objection concerned with reality. The concept of a particle, for example, a grain of sand, contains implicitly the notion that it is at a definite position and has a definite motion. But according to quantum mechanics it is impossible to determine simultaneously with arbitrary accuracy position and motion (more correctly momentum, that is, mass times velocity). Thus two questions arise. First, what is there to prevent us from measuring both quantities with arbitrary accuracy by refined experiments, in spite of the theoretical assertion? Second, if it should really turn out that this is not feasible, are we still justified in applying to the electron the concept of particle and the ideas associated with it?

With regard to the first question, it is clear that if the theory is correct—and we have sufficient grounds for believing this—the obstacle to simultaneous measurability of position and motion (and of other similar pairs of so-called 'conjugate' quantities) must lie in the laws of quantum mechanics itself. This is indeed the case, but it is not at all obvious. NIELS BOHR himself has devoted much labour and ingenuity to developing a theory of measurements to clear up this situation and to meet the most subtle considerations of EINSTEIN, who repeatedly tried to think out measuring devices by means of which position and motion could be measured simultaneously and exactly. The conclusion is as follows. In order to measure space coordinates and instants of time rigid measuring rods and clocks are required. On the other hand to measure momenta and energies arrangements with movable parts are needed to take up and indicate the impact of the object to be measured. If we take into consideration the fact that quantum mechanics is appropriate for dealing with the interaction of object and apparatus, we see that no arrangement is possible that satisfies both conditions at the same time. There exist, therefore, mutually exclusive but complementary experiments, which only in combination with each other disclose all that can be learned about an object. This idea of complementarity in physics is generally regarded as the key to the intuitive understanding of quantum processes. BOHR has transferred the idea in an ingenious manner to completely different fields—for example, to the relationship between consciousness and brain, to the problem of free will, and to other fundamental problems of philosophy.

Now to come to the final point—can we still call something with which the concepts of position and motion cannot be associated in the usual way a *thing*, a *particle*? And if not, what is the reality that our theory has been invented to describe?

The answer to this question is no longer physics, but philosophy, and to deal with it completely would overstep the bounds of this lecture. I have expounded my views on it fully elsewhere [23]. Here I will only say that I am emphatically for the retention of the particle idea. Naturally it is necessary to redefine what is meant. For this purpose well-developed concepts are available, which are familiar in mathematics under the name of invariants with respect to transformations. Every object that we perceive appears in innumerable aspects. The concept of the object is the invariant of all these aspects. From this point of view, the present universally used conceptual system, in which particles and waves occur at the same time, can be completely justified.

The most recent research on nuclei and elementary particles has, however, led us to limits beyond which this conceptual system in its turn does not appear to suffice. The lesson to be learned from the story I have told of the origin of quantum mechanics is that, presumably, a refinement of mathematical methods will not suffice to produce a satisfactory theory, but that somewhere in our doctrine there lurks a concept not justified by any experience, which will have to be eliminated in order to clear the way.

REFERENCES

1 R. LADENBURG, Z. Physik 4, 451 (1921); R. LADENBURG and F. REICHE, Naturwiss. 11, 584 (1923).

2 H. A. KRAMERS, Nature 113, 673 (1924).

3 —— and W. HEISENBERG, Z. Physik 31, 631 (1925).

4 M. BORN, ibid. 26, 379 (1924); M. BORN and P. JORDAN, ibid. 33, 479 (1925).

5 W. HEISENBERG, ibid. 33, 879 (1925).

6 M. BORN and P. JORDAN, ibid. 34, 358 (1925).

7 M. BORN, W. HEISENBERG, P. JORDAN, ibid. 35, 557 (1926).

8 P. A. M. DIRAC, Proc. Roy. Soc. (London) A109, 642 (1925).

9 W. PAULI, Z. Physik 36, 336 (1926).

10 E. SCHRÖDINGER, Ann. Physik (4), 79, 361, 489, 734 (1926); 80, 437 (1926); 81, 109 (1926).

11 LOUIS DE BROGLIE, Thèses, Paris, 1924; Ann. Physik (10), 3, 22 (1925).

12 W. ELSASSER, Naturwiss. 13, 711 (1925).

13 C. J. DAVISSON and L. H. GERMER, Phys. Rev. 30, 707 (1927).

14 G. P. THOMSON and A. REID, Nature 119, 890 (1927); G. P. THOMSON, Proc. Roy. Soc. (London) A117, 600 (1928).

15 E. SCHRÖDINGER, Brit. J. Phil. Sci. 3, 109, 233 (1952).

16 M. BORN and N. WIENER, Z. Physik 36, 174 (1926).

17 M. BORN, ibid. 37, 863 (1926); 38, 803 (1926); Gött. Nachr. Math.-Physik K1, 1, 146 (1926).

18 G. WENTZEL, Z. Physik 40, 590 (1926).

19 W. HEISENBERG, ibid. 43, 172 (1927).

20 M. BORN, ibid. 40, 167 (1926); M. BORN and V. FOCK, ibid., 51, 165 (1928).

21 P. A. M. DIRAC, Proc. Roy. Soc. (London) A109, 642 (1925); 110, 561 (1926); 111, 281 (1926); 112, 674 (1926).

22 NIELS BOHR, Naturwiss. 16, 245 (1928); 17, 483 (1929); 21, 13 (1933); 'Causality and complementarity,' Die Erkenntnis 6, 293 (1936).

23 M. BORN, Phil. Quart. 3, 134 (1953); Physik. Bl. 10, 49 (1954).

PHYSICS AND RELATIVITY

[A lecture given at the International Relativity Conference in Berne, Switzerland, on July 16, 1955.]

I have been honoured by being asked to give the address on Physics and Relativity in place of NIELS BOHR who was prevented from coming to Berne.

I do not know what BOHR had in mind when he chose the title. I cannot remember that I have ever discussed relativity with him; there was in fact nothing to discuss as we agreed on all essential points. The title Physics and Relativity may be interpreted in different ways: it may mean either a review of the empirical facts on which relativity was built, or it may mean a survey of the consequences of relativity for the whole of physics. Now such a survey was just the purpose of this conference, and it would be presumptuous and quite beyond my power to summarize all the reports and investigations. I propose instead to give you an impression of the situation of physics 50 years ago when EINSTEIN's first papers appeared, to analyze the contents of these papers in comparison with the work of his predecessors and to describe the impact of them on the world of physics. For most of you this is history. Relativity was an established theory when you began to study. There are very few left who, like me, can remember those distant days. For my contemporaries EINSTEIN's theory was new and revolutionary, an effort was needed to assimilate it. Not everybody was able or willing to do so. Thus the period after EINSTEIN's discovery was full of controversy, sometimes of bitter strife. I shall try to revive these exciting days when the foundation of modern physics was laid, by telling the story as it appeared to me.

When I began to study in the year 1901 MAXWELL's theory was accepted everywhere but not taught everywhere. A lecture by CLEMENS SCHAEFER which I attended at Breslau University was the first of its kind there and appeared to us to be very difficult. When I came to Göttingen in 1904 I attended a lecture on optics by WOLDEMAR VOIGT, which was based on MAXWELL's theory; but that was a new venture, the transition from the elastic ether theory was only a few years old. The main representative of the modern spirit in theoretical physics at Göttingen was at that time MAX ABRAHAM, whose well-known book, then called *Abraham-Föppl*, now *Abraham-Becker*, was our main source of information. All this is to indicate the scientific atmosphere in which we grew up. NEWTON's mechanics still dominated the field completely, in spite of the revolutionary discoveries made during the preceding decade, X-rays, radioactivity, the electron, the radiation formula and the quantum of energy, etc. The student was still taught—and I think not only in Germany, but everywhere—that the aim of physics was to reduce all phenomena to the motion of particles according to NEWTON's laws, and to doubt these laws was heresy never attempted.

My first encounter with the difficulties of this orthodox creed happened in 1905, the year which we celebrate today, in a seminar on the theory of electrons, held not by a physicist but by a mathematician, HERMANN MINKOWSKI. My memory of these long bygone days is of course blurred, but I am sure that in this seminar we discussed what was known at this period about the electrodynamics and optics of moving systems. We studied papers by HERTZ, FITZGERALD, LARMOR, LORENTZ, POINCARÉ, and others but also got an inkling of MINKOWSKI's own ideas which were published only two years later.

I have now to say some words about the work of these predecessors of EINSTEIN, mainly of LORENTZ and POINCARÉ. But I confess that I have not read again all their innumerable papers and books. When I retired from my chair at Edinburgh I settled at a quiet place where no scientific library is available, and I got rid of most of my own books. Therefore I rely a good deal on my own memory, assisted by a few books which I shall quote.

H. A. LORENTZ' important papers of 1892 and 1895 on the electrodynamics of moving bodies contain much of the formalism of relativity. However, his fundamental assumptions were quite unrelativistic. He assumed an ether absolutely at rest, a kind of materialization of NEWTON's absolute space, and he also took NEWTON's absolute time for granted. When he discovered that his field equations for empty space were invariant for certain linear transformations, by which the co-ordinates x, y, z and the time t were simultaneously transformed into new parameters x', y', z', t', he called them 'local co-ordinates' and 'local time.' These transformations, for which POINCARÉ later introduced the term Lorentz transformations, were in fact older; already in 1887 W. VOIGT had observed that the wave equation of the elastic theory of light was invariant with respect to this type of transformations. LORENTZ has further shown that if the interaction of matter and light was regarded to be due to electrons imbedded in the substance all observations concerning effects of the first order in $\beta = v/c$ (v=velocity of matter, c=velocity of light) could be explained, in particular the fact that no first order effect of the movement of matter could be discovered by an observer taking part in the motion. But there were some very accurate experiments such as that performed by MICHELSON first in 1881 in Potsdam, and repeated with higher accuracy in America in 1887 by MICHELSON and MORLEY, which showed that no effect of the earth motion could be found even to the second order in β. To explain this FITZGERALD invented in 1892 the contraction hypothesis, which was at once taken up by LORENTZ and included in his system. Thus

LORENTZ obtained a set of field equations for moving bodies which was in agreement
with all known observations; it was relativistic invariant for processes in empty space,
and approximately invariant (up to terms of 1st order in β) for material bodies. Still
LORENTZ stuck to his ether at rest and the traditional absolute time. I shall return
to this point presently. When HENRI POINCARÉ took up this investigation, he went
a step further. In regard to his work I refer to the excellent book by Sir EDMUND
WHITTAKER, *A History of the Theories of Aether and Electricity,* which was already
in use as a guide in my student times. It has now been completely rewritten. The
second volume of the new edition deals with 'The Modern Theories, 1900–1926';
there you can find quotations from POINTCARÉ's papers, some of which I have
looked up in the original. They show that as early as 1899 he regarded it as very
probable that absolute motion is indetectable in principle and that no ether exists.
He formulated the same ideas in a more precise form, though without any mathematics, in a lecture given in 1904 to a Congress of Arts and Science at St. Louis,
U.S.A., and he predicted the rise of a new mechanics which will be characterized
above all by the rule, that no velocity can exceed the velocity of light.

WHITTAKER was so impressed by these statements that he gave to the relevant
chapter in his book the title 'The Relativity Theory of Poincaré and Lorentz.'
EINSTEIN's contributions appear there as being of minor importance.

I have tried to form an opinion about this question from my own recollections
and with the help of a few publications available to me.

In the happy years before the first World War the Academy of Göttingen had
a considerable fund, called the Wolfskehl-Stiftung (W.-Foundation) which was
given originally with the direction to award a prize of 100,000 Marks for the proof
of FERMAT's celebrated 'Great Theorem.' Hundreds of letters, or even just post-
cards, arrived every year claiming to contain the solution, and the mathematicians
were kept busy to discover the error. The futility of this process became so annoying
that it was decided to use the money for other more useful purposes, namely to
invite distinguished scholars to lecture on current scientific problems. One of these
series of lectures was given by HENRI POINCARÉ, April 22nd-28th 1909, and has
been published as a book by Teubner in 1910. I have attended these POINCARÉ-
Festspiele (P.-Festival), as we called it, and now refreshed my memory by looking
through the book. The first five lectures dealt with purely mathematical problems;
the sixth lecture had the title 'La mécanique nouvelle.' It is a popular account of
the theory of relativity without any formulae and with very few quotations. EINSTEIN
and MINKOWSKI are not mentioned at all, only MICHELSON, ABRAHAM and LORENTZ.
But the reasoning used by POINCARÉ was just that, which EINSTEIN introduced in
his first paper of 1905, of which I shall speak presently. Does this mean that POIN-
CARÉ knew all this before EINSTEIN? It is possible, but the strange thing is that this

lecture definitely gives you the impression that he is recording Lorentz' work.

On the other hand Lorentz himself has never claimed to be the author of the principle of relativity. The year after Poincaré's visit to Göttingen we had the Lorentz-Festspiele. I, at the time a young Privatdocent, was appointed temporary assistant to the distinguished guest and charged with taking notes of the lectures and preparing them for publication. Thus I was privileged with having daily discussions with Lorentz. The lectures have appeared in *Physikalische Zeitschrift* (vol. 11, 1910, p. 1234). The second lecture begins with the words: 'Das Einsteinsche Relativitätsprinzip hier in Göttingen zu besprechen, wo Minkowski gewirkt hat, erscheint mir eine besonders willkommene Aufgabe.' 'To discuss Einstein's Principle of Relativity here in Göttingen where Minkowski has taught seems to me a particularly welcome task.' This suffices to show that Lorentz himself regarded Einstein as the discoverer of the principle of relativity. On the same page and also in the following sections are other remarks which reveal Lorentz' reluctance to abandon the ideas of absolute space and time. When I visited Lorentz a few years before his death, his skepticism had not changed.

I have told you all these details because they illuminate the scientific scene of 50 years ago, not because I think that the question of priority is of great importance.

May I now return to my own struggle with the relativity problem. After having graduated Dr. phil. in Göttingen I went in 1907 to Cambridge to learn something about the electron at the source. J. J. Thomson's lectures were very stimulating indeed; he showed brilliant experiments. But Larmor's theoretical course did not help me very much; I found it very hard to understand his Irish dialect, and what I understood seemed to me not on the level of Minkowski's ideas. I then returned to my home city Breslau, and there at last I heard the name of Einstein and read his papers. I was working at that time on a relativistic problem, which was an offspring of Minkowski's seminar, and talked about it to my friends. One of them, Stanislaus Loria, a young Pole, directed my attention to Einstein's articles, and thus I read them. Although I was quite familiar with the relativistic idea and the Lorentz transformations, Einstein's reasoning was a revelation to me.

Many of you may have looked up his paper 'Zur Elektrodynamik bewegter Körper' in *Annalen der Physik* (4), vol. 17, p. 811, 1905, and you will have noticed some peculiarities. The striking point is that it contains not a single reference to previous literature. It gives you the impression of quite a new venture. But that is, of course, as I have tried to explain, not true. We have Einstein's own testimony. Dr. Carl Seelig, who has published a most charming book on *Einstein und die Schweiz* asked Einstein which scientific literature had contributed most to his ideas on relativity during his period in Bern, and received an answer on February 19th of this year which he published in the *Technische Rundschau* (N. 20, 47. Jahrgang, Bern 6. Mai

1955); EINSTEIN wrote:

'Es ist zweifellos, daß die spezielle Relativitätstheorie, wenn wir ihre Entwicklung rückschauend betrachten, im Jahre 1905 reif zur Entdeckung war. LORENTZ hatte schon erkannt, daß für die Analyse der MAXWELLschen Gleichungen die später nach ihm benannte Transformation wesentlich sei, und POINCARÉ hat diese Erkenntnis noch vertieft. Was mich betrifft, so kannte ich nur LORENTZ bedeutendes Werk von 1895—"La théorie électromagnétique de MAXWELL" und "Versuch einer Theorie der elektrischen und optischen Erscheinungenin bewegten Körpern"—aber nicht LORENTZ', spätere Arbeiten, und auch nicht die daran anschließende Untersuchung von POINCARÉ. In diesem Sinne war meine Arbeit von 1905 selbständig.

'Was dabei neu war, war die Erkenntnis, daß die Bedeutung der Lorentztransformation über den Zusammenhang mit den MAXWELLschen Gleichungen hinausging und das Wesen von Raum und Zeit im allgemeinen betraf. Auch war die Einsicht neu, daß die "Lorentz-Invarianz" eine allgemeine Bedingung sei für jede physikalische Theorie. Das war für mich von besonderer Wichtigkeit, weil ich schon früher erkannt hatte, daß die MAXWELLsche Theorie die Mikrostruktur der Strahlung nicht darstelle und deshalb nicht allgemein haltbar sei—.'

Translated:

'There is no doubt, that the special theory of relativity, if we regard its development in retrospect, was ripe for discovery in 1905. LORENTZ had already observed that for the analysis of MAXWELL's equations the transformations which later were known by his name are essential, and POINCARÉ had even penetrated deeper into these connections. Concerning myself, I knew only LORENTZ' important work of 1895 (the two papers quoted above in the German text) but not LORENTZ' later work, nor the consecutive investigations by POINCARÉ. In this sense my work of 1905 was independent. The new feature of it was the realization of the fact that the bearing of the LORENTZ transformation transcended its connection with MAXWELL's equations and was concerned with the nature of space and time in general. A further new result was that the "Lorentz invariance" is a general condition for any physical theory. This was for me of particular importance because I had already previously found that MAXWELL's theory did not account for the micro-structure of radiation and could therefore have no general validity—.'

This, I think, makes the situation perfectly clear. The last sentence of this letter is of particular importance. For it shows that EINSTEIN's papers of 1905 on relativity and on the light quantum were not disconnected. He believed already then that MAXWELL's equations were only approximately true, that the actual behaviour of light was more complicated and ought to be described in terms of light quanta (or photons, as we say today), but that the principle of relativity was more general and should be founded on considerations which would be still valid when MAXWELL's equations had to be discarded and replaced by a new theory of the fine structure of light (our present quantum electrodynamics).

The second peculiar feature of this first relativity paper by EINSTEIN is his point of departure, the empirical facts on which he built his theory. It is of surprising simplicity. He says that the usual formulation of the law of induction contains an asymmetry which is artificial, and does not correspond to facts. According to observation, the current induced depends only on the relative motion of the conducting wire and the magnet, while the usual theory explains the effect in quite different terms according to whether the wire is at rest and the magnet moving or vice versa. Then there follows a short sentence referring to the fact that all attempts to discover experimentally the movement of the earth through the ether have failed. It gives you the impression that MICHELSON's experiment was not so important after all, and that EINSTEIN would have arrived at his relativity principle in any case.

This principle together with the postulate that the velocity of light is constant, independent of the system of reference, are the only assumptions from which the whole theory is derived on a few pages. The first step is the demonstration that absolute simultaneity of two events at different places has no physical meaning. Then relative simultaneity is defined by setting the clocks at different places in a system of reference in such a way that a light signal needs the same time either way between two of them. This definition leads directly to the Lorentz transformations and all their consequences: the Lorentz-Fitzgerald contraction, the time dilation, the addition theorem of velocities, the transformation law for the electromagnetic field components in vacuum, the Doppler principle, the aberration effect, the transformation law for energy, the equations of motion for an electron and the formulae for the longitudinal and transversal mass as functions of the velocity.

But for me—and many others—the exciting feature of this paper was not so much its simplicity and completeness, but the audacity of challenging ISAAC NEWTON's established philosophy, the traditional concepts of space and time. That distinguishes EINSTEIN's work from his predecessors and gives us the right to speak of EINSTEIN's theory of relativity, in spite of WHITTAKER's different opinion.

EINSTEIN's second paper on relativity 'Ist die Trägheit eines Körpers von seinem Energieinhalt abhängig?' (*Ann. d. Phys.* (4), vol. 18, 1905, p. 639) contains on three pages a proof of the celebrated formula $E=mc^2$ expressing the equivalence of mass and energy, which has turned out to be of fundamental importance in nuclear physics, for the understanding of the structure of matter and of the source of stellar energy as well, and for the technical exploitation of nuclear energy, for bad or good. This paper also has become the object of priority disputes. In fact, the formula had been known for special cases; for instance the Austrian physicist F. HASENÖHRL had shown already in 1904 that electromagnetic radiation enclosed in a vessel produced an increase of its resistance to acceleration, i.e., its mass, proportional to the radiation energy. HASENÖHRL was killed in the first world war and could not object when his name was later misused to discredit EINSTEIN's discovery. However, I shall not enter into an account of this sordid story. I have mentioned these matters only to make it clear that special relativity was, after all, not a one-man discovery. EINSTEIN's work was the keystone to an arch which LORENTZ, POINCARÉ and others had built and which was to carry the structure erected by MINKOWSKI. I think it wrong to forget these other men, as it can be found in many books. Even PHILIPP FRANK's excellent biography *Einstein, Sein Leben und seine Zeit*, cannot be acquitted of this reproach, e.g., when he says (in Chap. 3, No. 6 of the German edition) that nobody before EINSTEIN had ever considered a new type of mechanical law in which the velocity of light plays a prominent part. Both POINCARÉ and LORENTZ have been aware of this, and the relativistic expression for the mass (which contains c) has rightly been called LORENTZ' formula.

Today this formula is taken so much for granted that you can hardly imagine the acerbity of the controversies which raged around it. In 1901 W. KAUFMANN in Göttingen had by an investigation of the electromagnetic deflection of fast cathode rays first established the fact that the mass of the electron depends on its velocity. MAX ABRAHAM, whom I have mentioned already, took up this challenge and showed that the electromagnetic mass, as introduced by J. J. THOMSON, i.e., the self-energy of the electron's own field, properly developed for high velocities did indeed depend on velocity. He assumed the electron to be a rigid sphere; but later he also modified his theory by taking account of the Lorentz-Fitzgerald contraction, and obtained exactly the formula which LORENTZ had already found by a simpler reasoning. As a matter of fact, the velocity dependence of energy and of mass has nothing at all to do with the structure of the body considered, but is a general relativistic effect. Before this became clear, many theoreticians wrote voluminous, not to say monstrous, papers on the electromagnetic self-energy of the rigid electron—G. HERGLOTZ, P. HERTZ, A. SOMMERFELD, and others. My first scientific attempt was also in this direction; however, I did not assume the electron to be rigid in the classical sense, but tried to define relativistic rigidity by generalizing the Lorentz electron for accelerated motion, with the help of the methods I had learned from MINKOWSKI.

Today all these efforts appear rather wasted; quantum theory has shifted the point of view, and at present the tendency is to circumvent the problem of self-energy rather than to solve it. But one day it will return to the center of the scene.

MINKOWSKI published his paper 'Die Grundlagen für die elektromagnetischen Vorgänge in bewegten Körpern' in 1907. It contained the systematic presentation of his formal unification of space and time into a four-dimensional 'world' with a pseudo-euclidean geometry, for which a vector- and tensor calculus is developed. This calculus, with some modifications, soon became the standard method of all relativistic investigations. Moreover, MINKOWSKI's paper contained important new results: a set of equations for the electromagnetic field in moving material bodies which is exactly invariant with respect to LORENTZ transformation, not only a first approximation as LORENTZ' slightly different equations; further a new approach to the mechanical equations of motion.

In the beginning of 1908 I had the audacity to send my manuscript on the electron to MINKOWSKI, and he was kind enough to answer. On September 21st of the same year I listened at Cologne to his famous lecture 'Raum und Zeit,' in which he explained his ideas in popular form to the members of the Naturforscher-Versammlung. He invited me to come to Göttingen and to join him in further work. So I did; but alas, after a few weeks our collaboration ended through MINKOWSKI's sudden death. It fell to me to sift his unpublished papers, one of which I succeeded to reconstruct and to publish.

My first meeting with EINSTEIN happened in the following year, 1909, at the Naturforscher-Versammlung in Salzburg. There EINSTEIN gave a lecture with the title 'Über die neueren Umwandlungen, welche unsere Anschauungen über die Natur des Lichtes erfahren haben,' which means obviously the introduction of the light quantum. I also gave a talk 'Die Dynamik des Elektrons im System des Relativitätsprinzips.' This seems to me rather amusing: EINSTEIN had already proceeded beyond special relativity which he left to minor prophets, while he himself pondered about the new riddles arising from the quantum structure of light, and of course about gravitation and general relativity which at that time was not ripe for general discussion.

From this time on I saw EINSTEIN occasionally at conferences and exchanged a few letters with him. He became professor at the University of Zürich in 1909, then at Prague in 1910 and returned to Zürich, as professor at the Polytechnicum in 1912. Already in the following year he went to Berlin, where the Prussian Academy had offered him a special chair, vacated by the death of VAN'T HOFF, with no teaching obligations, and with other privileges. This invitation was mainly due to the efforts of MAX PLANCK who was deeply interested in relativity and had contributed important papers on relativistic mechanics and thermodynamics. Two years later, in spring 1915, I was also called to Berlin by PLANCK, to assist him in his teaching. The following four years have been amongst the most memorable of my life, not because the first World War was raging with all its sorrows, excitements, privations and indignities, but because I was near to PLANCK and EINSTEIN.

It was the only period when I saw EINSTEIN very frequently, at times almost daily, and when I could watch the working of his mind and learn his ideas on physics and on many other subjects.

It was the time when general relativity was finally formulated. Now this was, in contrast to the special theory, a real one-man work. It began with a paper published as early as December, 1907, which contains the principle of equivalence, the only empirical pillar on which the whole imposing structure of general relativity was built.

When speaking of the physical facts which EINSTEIN used in 1905 for his special relativity I said that it was the law of electromagnetic induction which seemed to have guided EINSTEIN more than even MICHELSON's experiment. Now the induction law was at that time about 70 years old (FARADAY discovered it in 1834), everybody had known all along that the effect depended only on relative motion, but nobody had taken offence at the theory not accounting for this circumstance.

Now the case of the equivalence principle is very similar, only that the critical empirical fact has been known by everybody far longer, namely about 250 years. GALILEO had found that all bodies move with the same acceleration under terrestrial gravity, and NEWTON generalized this for the mutual gravitational attraction of celestial bodies. This fact, namely, that the inertial and the gravitational mass are equal, was taken as a peculiar property of NEWTON's force, and nobody seems to have pondered about it.

Special relativity had restored the special rôle and the equivalence of the inertial systems of Newtonian mechanics for the whole of physics; absolute motion was indetectable as long as no accelerations occurred. But the inertia effects, the centrifugal forces and corresponding electromagnetic phenomena, which appear in accelerated, for instance rotating, systems could be described only in terms of absolute space. This seemed to be intolerable to EINSTEIN. Brooding over it, he noticed that the equality of inertial and gravitational mass implied that an observer in a closed box could not decide whether a non-uniformity of the motion of a body in the box was due to an acceleration of the whole box or to an external gravitational field. This gave him the clue for general relativity. EINSTEIN postulated that this equivalence should hold as a general principle for all natural phenomena, not only mechanical motion. Thus he arrived in 1911 at the conclusion that a beam of light must be bent in a gravitational field and suggested at once that his simple formula of deflection could be experimentally checked by observing the position of fixed stars near the sun during a total eclipse.

The actual development of the theory was a tremendous task, for a new branch of mathematics, quite unfamiliar to physicists, had to be used. Some more conservative physicists, ABRAHAM, MIE, NORDSTRÖM and others tried to develop from EINSTEIN's equivalence principle a coherent scalar theory of the gravitational field, with little success. EINSTEIN himself was the only one who discovered the right mathematical tool in RIEMANN's geometry, as extended by RICCI and LEVI-CIVITÀ, and he found in his old friend MARCEL GROSSMANN a skillful collaborator. But it took several years, until 1915, to finish this work.

I remember that on my honeymoon in 1913 I had in my luggage some reprints of EINSTEIN's papers which absorbed my attention for hours, much to the annoyance of my bride. These papers seemed to me fascinating, but difficult and almost frightening. When I met EINSTEIN in Berlin in 1915 the theory was much improved and crowned by the explanation of the anomaly of the perihelion of Mercury, discovered by LEVERRIER. I learned it not only from the publications but from numerous discussions with EINSTEIN—which had the effect that I decided never to attempt any work in this field. The foundation of general relativity appeared to me then, and it still does, the greatest feat of human thinking about Nature, the most amazing combination of philosophical penetration, physical intuition and mathematical skill. But its connections with experience were slender. It appealed to me like a great work of art, to be enjoyed and admired from a distance.

According to my interpretation of the title of this lecture I shall not enter into a discussion of the empirical confirmation of the special and the general theory of relativity, as I am no expert, and as others have spoken of it already I shall only just mention the most striking events.

In 1915 SOMMERFELD's relativistic theory of the fine structure of the hydrogen lines was published. It is based on the mathematical result, that the dependence of mass on velocity produces a precession of the perihelion of the elliptic orbit. It is quite interesting that POINCARÉ had already considered this effect to explain LEVERRIER's anomaly in the motion of the planet Mercury; a remark about this is contained in POINCARÉ's lecture in Göttingen quoted before. The result was of course negative, as the velocity of Mercury is much too small compared with that of light. It is different with the electron moving around a nucleus and, in conbination with the quantization laws of BOHR and SOMMERFELD, this led to the explanation of the splitting of the hydrogen lines.

The modern version of the theory of the hydrogen spectrum is based on DIRAC's relativistic wave equation and has recently been much refined with the help of quantum electrodynamics.

Another striking result of relativity combined with EINSTEIN's idea of light quanta is the theory of the Compton effect.

The time dilation effect was directly confirmed as the transversal DOPPLER effect on hydrogen canal rays in 1938 by IVES and STIEVELL, and with higher accuracy in 1939 by RÜCHARDT and OTTING. It plays an important part in the modern research on mesons in cosmic rays where the observed lifetime of a meson may be a hundred times as large as the intrinsic one in consequence of the large velocities.

At present special relativity is taken for granted, the whole of atomic physics is so merged with it, so soaked in it, that it would be quite meaningless to pick out particular effects as confirmations of EINSTEIN's theory. The situation in general relativity is different; all the three effects predicted by EINSTEIN exist, but the question of quantitative agreement between the theory and observation is still under discussion. However, the importance of general relativity lies in the revolution which it has produced in cosmology. It started in 1917 when EINSTEIN generalized his field equations by adding the so-called cosmological term and showed that a solution exists representing a closed universe. This suggestion of a finite, but unbounded, space is one of the very greatest ideas about the nature of the world which ever has been conceived. It solved the mysterious fact why the system of stars did not disperse and thin out, which it would do if space were infinite; it gave a physical meaning to MACH's principle which postulated that the law of inertia should not be regarded as a property of empty space but as an effect of the total system of stars, and it opened the way to the modern concept of the expanding universe. Here general relativity found again contact with observation through the work of the astronomers SHAPLEY, HUBBLE and many others. Today cosmology is an extensive science which has produced innumerable publications and books, of which I know little. Thus I am compelled to omit just that aspect of EINSTEIN's work which may be regarded as his greatest achievement.

May I, instead, tell you something about my personal relations with EINSTEIN in those bygone days and about the divergence of opinion which arose in the end between us in regard to the ultimate principles of physics.

The discussions which we had in Berlin ranged far beyond relativity, and even beyond physics at large. As the first world war was going on politics played of course a central part. But much as I would like to speak about these things I have to restrict myself to physics.

EINSTEIN was at that time working with DE HAAS on experiments about the so-called gyromagnetic effect, which proved the existence of AMPÈRE's molecular currents. He was also deeply interested in quantum theory but worried by its paradoxes.

In 1919 I became v. LAUE's successor at Frankfurt, and my companionship with EINSTEIN ceased. But we visited one another often and had a lively correspondence, of which I shall give you a few examples. It was the time when EINSTEIN suddenly became world famous, and his theory as well as his personality the object of fanatical controversy.

Just before the war a German expedition had gone to Russia to investigate EINSTEIN's prediction of the deflection of light by the sun during an eclipse; they were stopped by the outbreak of hostilities, and became prisoners of war. Now after the war two British expeditions went out for the same purpose, under the direction of Sir ARTHUR EDDINGTON, and they were successful. It is quite impossible to describe the stir which this event produced in the whole world. EINSTEIN became at once the most famous and popular figure, the man who had broken through the wall of hatred and united the scientists to a common effort, the man who had replaced ISAAC NEWTON's system of the world by another and better one. But at the same time an opposition, which had already been apparent while I was in Berlin, grew under the leadership of PHILIPP LENARD and JOHANNES STARK. It was springing from the most absurd mixture of scientific conservatism and prejudice with racial and political emotions, due to EINSTEIN's Jewish descent and pacifistic, antimilitaristic convictions. Here a few samples from EINSTEIN's letters; one of June 4th, 1919 begins with physics:

'. . . Die Quantentheorie löst bei mir ganz ähnliche Empfindungen aus wie bei Ihnen. Man müßte sich eigentlich der Erfolge schämen, weil sie nach dem jesuitischen Grundsatze gewonnen sind: "Die eine Hand darf nicht wissen, was die andre tut . . .".'

'. . . The quantum theory provokes in me quite similar sensations as in you. One ought really to be ashamed of the successes, as they are obtained with the help of the Jesuitic rule: "One hand must not know what the other does." '

and then, a few lines below, he continues about politics:

'. . . Darf ein hartgesottener X-Bruder und Determinist mit thränenfeuchten Augen sagen, daß er den Glauben an die Menschen verloren hat? Gerade des triebhafte Verhalten der Menschen von heute in politischen Dingen ist geeignet, den Glauben an den Determinismus recht lebendig zu machen . . . .'

'. . . Can a hardboiled X-brother (=mathematician; we used the expression "ixen," "to x," for "calculating") say with tears in his eyes that he has lost his faith in the human race? Just the instinctive behaviour of contemporary people in political affairs is apt to revive the belief in determinism . . . .'

You see that his deterministic philosophy which later created a gulf between him and the majority of physicists was not restricted to science but extended to human affairs as well.

At this time the inflation in Germany began to become serious. In my department STERN and GERLACH were preparing their well-known experiments, but hampered by the lack of funds. I decided to give a series of popular lectures on relativity with an entrance fee, using the general craze for information about this subject to raise funds for our researches. The plan was successful, the lectures were crowded, and when they appeared as a book three editions were quickly sold. EINSTEIN acknowledged my efforts by offering me the friendly 'Du' instead of the formal 'Sie' in a letter of November 9th, 1919, which also contains some suggestion how the Jews should react to the antisemitic drive going on:

'Also von jetzt ab soll Du gesagt werden unter uns, wenn Du es erlaubst . . . Ich würde es für vernünftig halten, wenn die Juden selbst Geld sammelten, um jüdischen Forschern außerhalb der Universitäten Unterstützung und Lehrgelegenheit zu bieten . . . .'

'Well, from now on the "Thou" shall be used between us, if thou agreest . . . I should think it reasonable if the Jews themselves would collect money in order to give Jewish scholars financial support and teaching facility outside the universities . . . .'

There appeared attacks against EINSTEIN by well-known scientists and philosophers in the *Frankfurter Zeitung* which aroused my pugnacity. I answered in a rather sharp article. EINSTEIN seems to have been pleased with it for he wrote on December 9th, 1919:

'Dein ausgezeichneter Artikel in der *Frankfurter Zeitung* hat mich sehr gefreut. Nun aber wirst Du, gerade wie ich, wenn auch in schwächerem Masstab, von Presse- und sonstigem Gelichter verfolgt. Bei mir ist es so arg, daß ich kaum mehr schnaufen, geschweige zu vernünftiger Arbeit kommen kann . . . .'

'Your excellent article in the *Frankfurter Zeitung* has given me great pleasure. Now you as well as I will be persecuted by gangs of pressmen and others though to a smaller degree. With me it is so bad that I can hardly breathe any more, to say nothing of doing reasonable work . . . .'

And about a year later (September 9th, 1920):

'. . . Wie bei dem Mann im Märchen alles zu Gold wurde, was er berührte, so wird bei mir alles zum Zeitungsgeschrei: Suum cuique . . . .'

'. . . Just as with the man in the fairy tale everything he touched was transformed into gold, with me everything becomes newspaper noise. Suum cuique . . . .'

If you are interested in that curious period when a whole world was excited about a physical theory which nobody understood, and when everywhere people were split into pro- and contra-EINSTEIN factions you can find an excellent account in the biography by PHILIPP FRANK quoted before.

However, scientific problems regained their proper place in our correspondence. In the same year (March 3rd, 1920) EINSTEIN wrote:

'Ich brüte in meiner freien Zeit immer über dem Quantenproblem vom Standpunkte der Relativität. Ich glaube nicht, daß die Theorie das Kontinuum wird entbehren können. Es will mir aber nicht gelingen, meiner Lieblingsidee, die Quantentheorie aus einer Uberbestimmung durch Differentialgleichungen zu verstehen, greifbare Gestalt zu geben . . . .'

'I always brood in my free time about the quantum problem from the standpoint of relativity. I do not think that the theory will have to discard the continuum. But I was unsuccessful, so far, to give tangible shape to my favourite idea, to understand the quantum theory with the help of differential equations by using conditions of over-determination . . . .'

Already at that time we discussed whether quantum theory could be reconciled with causality. Here a sentence from EINSTEIN's letter of January 27th 1920:

'. . . Das mit der Kausalität plagt mich auch viel. Ist die quantenhafte Licht-Absorption und -Emission wohl jemals im Sinne der vollständigen Kausalitätsforderung erfassbar oder bleibt ein statistischer Rest? Ich muss gestehen, dass mir da der Mut einer Überzeugung fehlt. Ich verzichte aber sehr, sehr ungern auf *vollständige* Kausalität . . . .'

'That question of causality worries me also a lot. Will the quantum absorption and emission of light ever be grasped in the sense of complete causality, or will there remain a statistical residue? I have to confess, that I lack the courage of a conviction. However I should be very, very loath to abandon *complete* causality . . . .'

From that time on our scientific ways parted more and more. I went to Göttingen and came in contact with NIELS BOHR, PAULI and HEISENBERG. When in 1927 quantum mechanics was developed, I hoped of course that EINSTEIN would agree, but was disappointed. Here a quotation from one of his letters (December 12th, 1926):

'. . . Die Quantenmechanik ist sehr achtunggebietend. Aber eine innere Stimme sagt mir, dass das doch nicht der wahre Jakob ist. Die Theorie liefert viel, aber dem Geheimnis des Alten bringt sie uns kaum näher. Jedenfalls bin ich überzeugt, daß der nicht würfelt . . . Ich plage mich damit herum, die Bewegungsgleichungen von als Singularitäten aufgefassten materiellen Punkten aus den Differentialgleichungen der allgemeinen Relativität abzuleiten . . . .'

'The quantum mechanics is very imposing. But an inner voice tells me that it is still not the true Jacob [a German colloquialism]. The theory yields much, but it hardly brings us nearer to the secret of the Old One. In any case I am convinced that he does not throw dice . . . I am toiling at deriving the equations of motion of material particles regarded as singularities from the differential equations of general relativity . . . .'

The last sentence refers to a paper which was finished much later at Princeton in collaboration with BENESH HOFFMANN and LEOPOLD INFELD, EINSTEIN's last great contribution to relativity. The assumption made in the original theory, that a free particle (e.g., a celestial body) moves on a geodesic turned out to be unnecessary, it could be derived from the field equations by a subtle procedure of successive approximations. These very deep and important investigations have been further developed by FOCK and INFELD.

The first part of the letter quoted refers to EINSTEIN's refusal to accept statistical laws in physics as final; he speaks of the diceplaying God, an expression which he has used later very often in discussion and letters.

During the last period of his life in Princeton he concentrated all his powers and energies on developing a new foundation of physics in conformity with his fundamental philosophical convictions, namely that it must be possible to think of the external world as existing independently of the observing subject, and that the laws governing this objective world are strictly causal, in the sense of deterministic. This was the aim of his unified field theories, of which he published several versions, always hoping that the quantum principles would in the end turn out to be a consequence of his field equations.

I cannot say much about these attempts, as right from the beginning I just did not believe in their success and therefore did not study his difficult papers with sufficient care. I think that quantum mechanics has followed up EINSTEIN's original philosophy, which led him to tremendous success, more closely than he did himself in his later period.

What is this lesson we learned from him? He himself has told us that he learned it from ERNST MACH, and therefore the positivists have claimed him to be one of them. I do not think this is true, if positivism is the doctrine that the purpose of science is the description of interrelation of sense impressions. EINSTEIN's leading principle was simply that something of which you could think and form a concept, but which from its very nature could not be submitted to an experimental test (like the simultaneity of events at distant places) has no physical meaning.

The quantum effects showed that this holds for a great many concepts of atomic physics, but EINSTEIN refused to apply his criterion to these cases. Thus he rejected the current interpretation of quantum mechanics, though it follows his own general teaching, and tried quite a different way, rather remote from experience. He had achieved his greatest success by relying on just *one* empirical fact known to every schoolboy. Yet now he tried to do without any empirical facts, by pure thinking. He believed in the power of reason to guess the laws according to which God has built the world. He was not alone in this conviction. One of the principal exponents of it was EDDINGTON in his later papers and books. In 1943 I published a pamphlet with the title *Experiment and Theory in Physics* (Cambridge University Press) in which I tried to analyze the situation and to refute EDDINGTON's claims. I sent a copy to EINSTEIN and received a very interesting reply which unfortunately has been lost; but I remember a phrase like this: 'Your thundering against the Hegelism is quite amusing, but I shall continue with my endeavours to guess God's ways.' A man of EINSTEIN's greatness who has achieved so much by thinking, has the right to go to the limit of the *a priori* method. Current physics has not followed him; it has continued to accumulate empirical facts, and to interpret them in a way which EINSTEIN thoroughly disliked. For him a potential or a field component was a real natural object which changed according to definite deterministic laws. Modern physics operates with wave functions which, in their mathematical behaviour, are very similar to classical potentials but do not represent real objects; they serve for determining the probability of finding real objects, whether these are particles or electromagnetic potentials, or other physical quantities. EINSTEIN made many attempts to prove the inconsistency of this theory with the help of ingenious examples and models, and NIELS BOHR took infinite trouble to refute these attacks; he has given a charming report about his discussions with EINSTEIN in the book *Einstein, Philosopher-Scientist* (The Library of Living Philosophers, Vol. 7, p. 199).

I saw EINSTEIN the last time about 1930, and although our correspondence continued I do not feel competent to speak about the last phase of EINSTEIN's life and work. I hope that Professor PAULI will tell us something about it. I conclude my address by apologizing that it was so long. But my friendship with EINSTEIN was one of the greatest experiences of my life, and 'Ex abundantia enim cordis os loquitur,' or in good Scots: 'Neirest the heart, neirest the mouth.'

DEVELOPMENT AND
ESSENCE OF THE
ATOMIC AGE

[Lecture given to a meeting of journalists held at the Protestant Theological Academy of Loccum Abbey, Niedersachsen, Germany, on March 18, 1955, and repeated at several other meetings during the summer of 1955.]

In following the invitation to speak about the Atomic Age, its development and essence, I do not think that I am intended to enlarge upon physical discoveries and their applications to technological and military ends, but rather on what appears to me the historical roots of these discoveries and their consequences upon the destiny of Man. But a scientist like myself has little time for historical studies; I have to rely on the fact that during my long life of more than 70 years I have witnessed a section of modern history and pondered about it. Moreover, I have read or at least scanned a few books which may be useful for my purpose. For instance, I remember from my student's days SPENGLER's *Decline of the West* (Untergang des Abend-landes). I have also read a little in ARNOLD TOYNBEE's great work, and listened to some of his Gifford Lectures given at Edinburgh a few years ago. I mention these two authors together because both share the opinion that there are regularities or even laws in human history which can be revealed by a comparative study of different groups of nations and civilizations. What I actually know of European history is essentially due to a book much used at British schools and elementary University courses because of its admirable style and clarity, H. A. L. FISHER's *A History of Europe*. His standpoint can be seen from quoting a few lines of his Preface.

One intellectual excitement has, however, been denied to me. Men wiser and more learned than I have discerned in history a plot, a rhythm, a predetermined pattern. These harmonies are concealed from me. I can see only one emergency following upon another as wave follows upon wave, only one great fact with respect to which, since it is unique, there can be no generalizations, only one safe rule for the historian: that he should recognize in the development of human destinies the play of the contingent and the unforeseen. This is not a doctrine of cynicism and despair. The fact of progress is written plain and large on the page of history; but progress is not a law of nature. The ground gained by one generation may be lost by the next. Thoughts of men may flow into the channels which lead to disaster and barbarism.

There are apparently two historical schools, one of which believes that the histor-ical course of events obeys laws and has a meaning, and another which denies this.

As a scientist I am accustomed to search for regularities and laws in natural phenomena. I beg your forebearance if I consider also the problem in hand from this standpoint, yet in quite a different manner from that used by the two historians mentioned.

The dawn of a new historical age, for instance the transition from antiquity to the medieval period, is obviously not noticed by those who are alive at that time. Everything goes on without break, the life of the son is not much different from that of the father. The division into periods and ages is an invention of the historians made for the purpose of finding their way in the chaos of events. Even the beginning of the scientific-technological period in which we are living was a slow process stretching over more than a hundred years and hardly noticed by the people of that day.

At present things appear to be different. During the time of a few years something new has arrived which is transforming our lives. This new feature includes simultaneously a horrible threat and a brilliant hope: the threat of self-destruction of the human race, the hope of earthly paradise. And this is not a revelation of religious prophets or of philosophical sages, but these two possibilities are presented to the human race for choosing by science, the most sober activity of the mind. The threat of destruction in particular is demonstrated by impressive examples—Hiroshima and Nagasaki—which should suffice to convince. But I wish to say right from the beginning that the atom bombs used there were children's toys compared with the thermo-nuclear weapons developed since. I myself have not taken any part in the development of this chapter of science—nuclear physics. But I know enough of it to say that it is not a question of a simple multiplication of destructive power, which would lead to the annihilation of a certain number of unfortunate people, while a much greater number of more fortunates would escape. It is a radical and sweeping change of the situation. Already today the stock of A-, H- and U-bombs in the United States and in Russia is probably sufficient to wipe out mutually all larger cities in both countries, and presumably in addition all remaining centers of civilization, since almost all countries are more or less attached to one of the giant powers. But much worse things are in preparation, perhaps already available for application: for instance the cobalt bomb which produces a radioactive dust spreading over wide areas and killing all living creatures therein. Particularly sinister are the aftereffects of radioactive radiation on generations unborn; mutations may be induced which lead to a degeneration of the human race. OTTO HAHN whose discovery of the fission of uranium has set in motion this development—without his participation and much against his wish—recently described the true aspect of the situation in a radio lecture which has been published and widely read; I need to add nothing to it. There he has also mentioned the useful applications of nuclear physics, namely, the generation of energy, the production of isotopes as instruments in medicine and technology, and so on. These may indeed become a blessing in future days, but only if these future days exist. We are standing at a crossroad as the human race has never met before on her way through the centuries.

This 'to be or not to be' is, however, only a symptom of a state of our mental development. We have to ask: what is the deeper cause of the dilemma in which man has been involved?

The fundamental fact is the discovery that the matter which we and all things around us are made of is not solid and indestructible, but unstable, an explosive. We are all sitting, in the true sense of the word, on a powder barrel. This barrel has, it is true, rather strong walls, and we needed a few thousand years to drill a hole into it. Today we have just got through and may at any moment blow ourselves sky-high with a match.

This dangerous situation is simply a matter of fact. I shall return to the scientific facts later and describe them in more learned terms. But first I want to discuss the question: would it not have been possible to let the barrel untouched and to sit peacefully upon it without caring about its content? Or, without the use of this metaphor: could the human race not live and flourish without investigating into the structure of matter and thus to conjure up the peril of self-destruction?

To answer this question one needs a definite philosophical outlook on history. I am hardly entitled to claim any knowledge in this field, yet, as I proposed before, permit me to try and tackle it with the methods of a scientist.

Then the situation appears like this. Man is often defined as the 'thinking animal.' His rise depends on his ability to collect experiences and to act accordingly. Single individuals or groups of such lead the way, others follow and learn. This was an anonymous process through centuries; we know nothing of the men of early ages who invented the first tools and weapons, who learned cattle breeding and agriculture, who developed the languages and the art of writing. But we may be sure that there was a permanent struggle between the minority of progressively-minded people and the conservative crowd, as we observe it since written documents exist. The total number of men is large and increases with each improvement of the conditions of life. If the percentage of the gifted remains roughly constant their absolute number grows in the same rate as the total number of men. Simultaneously with each technical invention the possibility of new combinations increases. Hence the situation is similar to that of the calculus of compound interest: if the interest is added to the capital this increases, and with it the next instalment of interest, hence again the capital, and so on *ad infinitum*. One has what the mathematicians call an exponential increase.

This is, of course, only correct for the average, it is a statistical law. I am convinced that the laws of statistics are valid in history just as for the game of roulette, or in atomic physics, in stellar astronomy, in genetics and so on—namely, in all cases where one has to do with large numbers. This may be taken as an interpretation of the meaning of the sentence from FISHER's *History of Europe,* quoted above: 'The fact of progress is written plain and large on the page of history.' But if he continues, 'but progress is not a law of nature' he appears to have applied an obsolete notion of the essence of natural laws, namely, that they are rigorously causal and deterministic and permit no exception. We know today that most of the laws of nature are of a statistical kind and permit deviations; we physicists call these 'fluctuations.'

As this idea is not familiar to everybody allow me to illustrate it by a simple example. The air which we all breathe seems to be a thin, continuous substance of uniform density. But investigations with intricate instruments have shown that actually the air consists of innumerable molecules (mainly of two kinds, oxygen and hydrogen) which fly about and collide with one another. The appearance of continuity is a consequence of the grossness of our senses which register only the average behaviour of big numbers of molecules. But then the question arises: why is the average distribution uniform in the chaotic dance of molecules? Or in other words, why is there the same number of molecules in two equal volumes of space? The answer is that there is never exactly the same number of molecules in equal volumes, but only approximately, and this is the consequence of a simple result of statistics, according to which this approximately uniform distribution has an overwhelming probability as compared with any others. But there are deviations which can be observed if the two volumes compared are sufficiently small. Particles suspended in the air, for instance pollen from plants or cigarette smoke, perform tiny irregular zig-zag motions which can be seen in a microscope; the explanation given by EINSTEIN of this effect, called Brownian movement, is simply that the number of air molecules hitting such a tiny, but microscopically visible, particle in opposite directions is not exactly equal in any short time interval, hence the particle is pushed about through the fluctuations of the average recoil. In principle there is no limit to the size of these fluctuations, but a statistical law makes it extremely improbable that very large deviations occur. Otherwise it might happen that the density of the air near to my mouth might become so small for a few minutes that I would suffocate. I am not afraid of this because the probability of its occurrence is immensely small.

I think that uniformity in history is due to the same statistical law. But ordinary history deals generally with small groups and short times; then the statistical uniformity does not strike the eye, but the fluctuations which appear chaotic and senseless. I wonder whether TOYNBEE's speculations may not be regarded as an attempt to discover regularities in the fluctuations.

However that may be, one conclusion from this consideration seems to me inescapable.

The process of gathering and applying knowledge seen as an endeavour of the whole human race over long periods of time must follow the statistical law of exponential increase and cannot be halted.

On the other hand, if only a restricted space on earth and a restricted period are considered, say a nation or a group of peoples in the period of a few hundred years, nothing of that process may be visible, even a loss of the achievements and a retrogression. But then the power of the human mind will manifest itself at another place of the world and at another time.

Let me illustrate this by a few historical reminiscences. The decisive step on the way to atomic physics was made about 2,500 years ago; I mean the speculations of the Greek school of natural philosophy, THALES, ANAXIMANDER, ANAXIMENES, especially the atomists LEUKIPPOS and DEMOCRITOS. They were the first who thought about Nature without expecting an immediate material advantage, driven by a pure desire of knowledge. They postulated the existence of natural laws and tried to reduce the variety of matter to the play of configuration and motion of invisible, unchangeable, equal particles. It is not easy to apprehend the immense superiority of this idea over all conceptions current at that time in the rest of the world. Together with the grand mathematics of the Greeks this idea might have led already at that early period to a decisive scientific-technological advance had not the social conditions been unfavourable. These Greek gentlemen lived in a world which venerated the harmony and beauty of body and mind. They despised manual work which was the task of slaves, and thus they neglected experiment which cannot be done without soiling one's hands. Thus no empirical foundation of the ideas was attempted, nor their technical application, which might have saved the antique world from the assault of the barbarians.

After the great migration of people the Christian Church erected a totalitarian system ill-disposed to all innovations. Yet the fire kindled by the Greeks smouldered under the ashes. It lay hidden in the books which were kept and copied in many monasteries and stored in the libraries of Byzantium, and it flared up to a bright flame through the Arabian scholars who even created essentially new things in mathematics and astronomy and who guarded the Greek tradition until the time was ripe. The Byzantians who fled before the Turks to Italy did bring with their books not only the knowledge of classical antiquity but also the idea of research. Thus came the time of discoveries and inventions which secured Europe's preponderance for a few centuries. A parallel development, perhaps of even older origin, took place in China. I know little about it, but there is a new comprehensive book by J. NEEDHAM, well known biologist at Cambridge, England, which gives a detailed account of it. During and after the European Renaissance, China was just in a state of rest or stagnation, and thus it came about that Europe was ahead for a few centuries. I had enough Chinese, also Japanese and Indian, students to be convinced that these nations are in no way inferior to us in scientific talent.

There are two conclusions to be drawn from these considerations. Firstly, it is quite absurd to believe the crisis in the existence of the human race, the dawn of the atomic age, might have been avoided, or the further development of dangerous knowledge might be inhibited. HITLER has tried to choke what he called 'Jewish Physics,' the Soviets tried the same with Mendelian genetics, both without any success, to their own detriment.

Secondly, the suddenness of the appearance of the critical situation is partly an historical accident, but mainly a deception of perspective distortion. The knowledge of Nature and the power springing from it are steadily growing, though with fluctuations and retrogressions, but in the average with the continuously increasing acceleration characterstic for a self-supporting (exponential) process. Thus the day had necessarily to come when the change of the conditions of life produced by this process would be considerable during one single generation and therefore would appear as a catastrophe. This impression of a catastrophe is increased by the complications due to the fact that there are peoples which have not taken part in the technical development and have to adapt themselves to it without proper preparation.

It is our generation which gathers the harvest sown by the Greek atomists. The final result of physical research is a confirmation of their fundamental idea that the material world is essentially composed of equal elementary particles whose configuration and interaction produces the variety of phenomena. But this simple description is, of course, only a crude condensation of an abundance of experimental results, and becomes, by supplementary features, in the end very complicated.

Those elementary particles are called nucleons, because by clotting together they form the atomic nucleus. The chemical atoms are neither invisible (as the name indicates), nor all identical for a definite chemical element, as believed during the last century. This is a consequence of the fact that a nucleon may be either electrically neutral—then it is called neutron—or may carry a positive elementary charge—then it is called proton. The chemical atoms consist of a nucleus which is extremely dense agglomeration of neutrons and protons (hence it is positively charged), and an extended cloud of negative electric particles, called electrons, surrounding the nucleus. The electron has a very small mass compared with the nucleon, but the same charge as the proton, with the opposite sign. The number of electrons in the cloud is equal to that of the protons in the nucleus so that the whole atom is electrically neutral. The electronic cloud determines the chemical and most of the physical properties of the atom. Atoms which have the same number of protons and therefore the same number of electrons in the cloud are chemically, and in most respects also physically, indistinguishable, even if the number of neutrons in the nucleus may differ. Such almost identical atoms, which differ only by the number of neutrons, i.e., by their mass (weight) are called isotopes.

The elements of ordinary chemistry and physics are mixtures of isotopes. The laws which govern the structure of the electronic cloud are known; the current research in this field is not concerned with the discovery of new principles but with the treatment of cases of increasing complexity. The laws governing the structure and behaviour of the nuclei are not so well explored. However, it is perfectly certain that some of the most general physical laws are valid there too, and with their help far-reaching conclusions can be drawn.

The most important of these laws is that which formulates the equivalence of mass ($\underline{M}$) and energy (E), expressed by the frequently quoted formula $E = \underline{M}c^2$ where c is the velocity of light. Its general derivation was given by EINSTEIN, exactly 50 years ago, with the help of relativistic reasoning, long before there existed any possibility of an experimental test. The number c is, in ordinary units, centimeters per second, very large, a 3 with 10 zeros behind it; hence $c^2 = c \times c$ is extremely large, a 9 with 20 zeros. Therefore the change of mass ($\underline{M} = E/c^2$) is excessively small for all ordinary chemical and physical energy exchanges. In principle a clock becomes a little heavier when wound up, but that is absolutely unmeasurable. The situation is different for nuclear transformations where large energies are exchanged.

A piece of a wall consisting of 100 equal bricks without mortar has a weight exactly 100 times that of a single brick; if there is mortar the weight is correspondingly higher. The same holds roughly for nucleons: a nucleus containing 100 nucleons is about 100 times as heavy as a single nucleon. Yet only approximately: there are deviations, hence there must be a kind of mortar. Now strangely enough this mortar appears to have a negative weight: the nucleus is lighter than the sum of its constituents. Namely, according to EINSTEIN, the mortar is the binding energy which is lost when the parts are combined. These 'mass-defects' are considerable, hence the corresponding energies enormously large.

The lightest element, hydrogen, consists of one isotope, the single proton. (There is also a hydrogen isotope with one additional neutron—deuteron—and one with two neutrons—triton.) The next element, helium, consists mainly of an isotope having 2 protons and 2 neutrons. When these agglomerate, a very great quantity of energy is liberated. The process does not occur spontaneously because there is an initial obstacle against the combination of the 4 particles, some energy has to be spent. The situation is like that of a water barrage, the gates of which have to be wound up before the water in the reservoir can stream out and do work. The same holds for the consecutive elements; they are instable and would combine unless there were barrages, fortunately very strong barrages, to keep them apart. This is the case in the series of elements up to the middle of the whole system, about the place of iron; from there on the situation is reversed, each nucleus has the tendency to split and is only prohibited to do so by a barrage. The last of the elements found in Nature, uranium, has the weakest barrage, and it was this one which was first broken in the experiments by HAHN and his collaborator STRASSMANN in 1938.

The way from these delicate laboratory experiments to the first uranium reactor (or pile) which was built in Chicago by ENRICO FERMI in 1942, was long and demanded an enormous amount of ingenuity, courage, skill, organization and money. The decisive discovery was that the fission of a uranium nucleus produced by the collision with a neutron is accompanied by the emission of several neutrons, and that the process could be directed in such a way that a sufficient number of these could be prevented from escaping or being lost by collisions with impurities as to produce an avalanche of new fissions, a self-containing reaction. To begin with, nobody could predict the outcome, but Nature has arranged it in this manner, hence it was discovered by Man as soon as the means were available. That they were available was an historical accident, a consequence of the great war. The technological process to produce a bomb until its explosion on July 16th, 1945, lasted three years and cost about half a billion dollars.

The inverse process, the fusion of nuclei into higher ones (e.g., hydrogen into helium) is the source of energy of the sun and of all stars. In the central parts of these the temperatures and pressures are so high that the combination of four nucleons is possible in a series of steps, through a chain reaction. The same has now been accomplished here on earth by using a uranium bomb as ignition. Thus we have now the H-bomb, which seemed to be an absolutely hellish invention, as no method of abating the violence of the explosion was known; but recently it has been announced that ways of controlling this reaction have been found.

There is no doubt any more: all matter is unstable. If this were not true the stars would not shine, there would be no heat and light from the sun, no life on earth. Stability and life are incompatible. Thus life is necessarily a dangerous adventure which may have a happy end or a bad one. Today the problem is how the greatest adventure of the human race can be directed towards a happy end.

Now I wish to say a few words about the blessings which can be obtained if men behave reasonably. There is, in the first line, the problem of energy. When I was young, half a century ago, the time our coal reserves would last was estimated to be a few hundred years; petrol oil was not used then on a large scale. Meanwhile an enormous amount of coal has been burnt, oil has been discovered and used in an ever increasing rate. Yet the estimate of the duration of the fossil fuel reserves is still many hundred years. Therefore it seems not to be an urgent problem to find new sources of energy. But this conclusion would be erroneous. Coal and oil are not only sources of energy but the most important raw materials for innumerable chemical products. May I just mention the plastics and their numerous applications. There will come a time when the agricultural output does not suffice for feeding the ever-increasing number of human beings. Then chemistry will be challenged to produce substitutes, for which, of course, coal is the only available raw material. Hence it is barbaric to burn coal and oil. Then the social aspect of the question must not be forgotten. The day seems to be not far away when in civilized countries no workmen will be available who are willing to take up the dark and dangerous profession of a miner, at least not for economically bearable wages. England seems to approach this state of affairs already. Then there are many countries which have neither coal nor oil; for these the easily transportable nuclear fuel will be a blessing.

Another type of the peaceful applications of nuclear physics are the radio-active by-products of atomic reactors. Instable, i.e., radio-active isotopes of many elements are produced, which can be applied to many purposes: as sources of radiation, instead of the expensive radium, in medicine, technology, agriculture; for instance for the treatment of cancer, the testing of materials, the production of new species of plants through mutations. Perhaps more important than all this is the idea of 'tracer elements.' By adding a small amount of a radio-active isotope to a given element it is possible to follow the fate of this element in chemical reactions, even in living organisms, by observing the radiation emitted. An ever increasing number of experiments in biological chemistry are already using this method, which marks a new epoch in our knowledge of the processes of life.

All this, and what may develop from it in days to come, are great things. An international conference at Geneva convened by UNO has discussed the exploitation of all these possibilities by a collaboration of all nations. I am not a nuclear physicist and have not attended it. I hope the labours of this meeting will bring in a rich harvest. But I cannot help asking: can even a technical paradise counterbalance the evil of the atomic bomb?

I have used the phrase 'paradise on earth' already in the beginning, but there I meant something different: not technical progress, but the realization of the eternal yearning of Man for 'peace on earth.'

In regard to the opinions I wish to express now, I cannot rely on my knowledge of physics, nor on my sporadic studies of history; they seem to me just common sense, and they are shared by a number of friends, leading scholars from different countries. We believe that a major war between Great Powers—there exist now only two or three—has become impossible, or at least will become impossible in the near future.

For it would lead, as I said already, in all likelihood to general destruction, not only of the fighting nations but also of the neutrals. CLAUSEWITZ' well known saying that war is the continuation of politics with other means does not hold any more, for war has become insanity, and if the human race is unable to renounce war its zoological name should not any longer be derived from sapientia but from dementia.

The leading statesmen seem to be well aware of this situation. The tuning down of the cold war which we are observing is an indication that it is so. The fear of the enormity of the catastrophe which might be the result of an armed conflict has led everywhere to approaches and negotiations. But fear is a bad foundation for reconciliation and solution of conflicts. Is it conceivable that the peace resting on fear which we very likely are attaining at present may be replaced by something better and more reliable?

I take it upon me if you regard me as a slightly ridiculous fellow who refuses to acknowledge an awkward situation

> Because, he argues trenchantly,
> What must not happen cannot be,

as the grotesque philosopher PALMSTRÖM says in the German poet MORGENSTERN's *Songs from the Gallows.**

However, I am not alone with this view. EINSTEIN shared it and has just before his death given a clear statement, together with BERTRAND RUSSELL, the great philosopher, and others. A number of 18 Nobel Laureates, chemists and physicists, gathered for a scientific discussion at Lindau, have unanimously accepted a declaration (the Mainau Statement) on similar lines. And many other people and groups of people have published similar declarations. May they appear today as dreamers: they are the builders of the future world.

But not much time is available for their words to take effect. All depends on this, the ability of our generation to readjust its thinking to the new facts. If it is unable to do so then the days of civilized life on earth are coming to an end. And even if all goes well, the way will pass very, very close to the abyss.

For the world is full of conflicts appearing insoluble: displaced frontiers of countries; expelled populations; antagonism of races, languages, national traditions, religions; the bankruptcy of the colonial system; and finally the opposing economical ideologies, capitalism and communism. Can we really hope that all these terrible tensions will be solved without application of force?

---

* CHRISTIAN MORGENSTERN wrote deep and beautiful poetry, which however found little resonance in the public. Then he published several little volumes of grotesque, apparently senseless verse under the title 'Galgenlieder' in which he caricatured his philosophy through two strange figures, PALMSTRÖM and KORFF. These books had a tremendous and lasting success.

Would it not be preferable, instead of the radical proposition to abandon war, to make an attempt to prohibit the new weapons of mass destruction by international agreement? This idea seems to me (and my friends) impracticable for the following reasons.

The production of energy through nuclear reactions is already being prepared and improved everywhere. A system of supervision intended to inhibit the production of weapons of destruction can function only in peace time. If war between major powers should break out which might initially be conducted with conventional weapons the supervision ceases. Is it reasonable to assume that a nation in distress but believing that she could save herself with the help of the atom bomb would be willing to renounce this last resource even if she is liable to suffer badly herself?

Concerning those 'conventional weapons' I must confess that I am unable to understand why they are not causing the same horror, the same detestation which is generally felt today towards the atomic weapons. They have ceased to be honest weapons used by soldiers against soldiers and have become means of indiscriminate destruction. They are not directed against military objects alone but against the whole organization and productive capacity of the enemy nation, against factories, railways, houses; they kill the helpless, the old, children, women, they destroy the most noble and valuable achievements of civilization, churches, schools, monuments, museums, libraries, without any regard for historical importance or irreparability. From the moral standpoint the decisive step of warfare towards modern barbarism was the concept of total war. Even without atomic weapons the prospect of the effects of using ordinary bombs, in combination with chemical and bacteriological poisons, is appalling enough.

Prohibition of atomic weapons alone is not justified, neither morally nor by the actual facts. The human race can only be saved by renouncing once for all the use of force through war. Today fear has produced such a precarious state of peace. The next aim must be to stabilize this peace by strengthening the ethical principles which secure the peaceful coexistence of men. CHRIST has taught how man ought to behave towards man. The nations have up to now acted—and the Churches have not objected to this attitude—as if these commandments are valid only inside their domains, but not in regard to their mutual relations. That is the root of the evil. We can only survive if in the international sphere distrust is replaced by understanding, jealousy by the will to help, hatred by love. In our time, before our eyes, the doctrine of non-violence has been victorious in the hands of a non-Christian, MAHATMA GHANDI, who has liberated his country without bloodshed (and I do not think that he would have acted differently if his adversaries had not been the well-meaning British, but any other nation). Why should it not be possible to follow his example?

I cannot make suggestions for the solution of the actual political conflict. Yet I wish to discuss a few general points.

The first of these is that a tremendous number of men in all countries have a personal interest in the preparation, and, if necessary, the waging of war. There are big industries and many types of business who make money from armaments. There are numerous men who like the life of a soldier because of its romantic tradition, or because they enjoy being rid of responsibility and having just to obey. There are the officers: generals, admirals, air marshals, etc., whose profession is war. They are still the advisers of present-day governments. Finally there are the physicists, chemists, engineers who invent new weapons and produce them. It would be an illusion to make an attempt of stabilizing the present precarious peace without taking any notice of all those people, without giving them some substitute for the loss they have to expect. How this can be done is beyond my competence except for one class of people, the physicists whom I know well. Here I see no great difficulty.

One hears often hard words about the atomic physicists: all calamity is the fault of these brain-athletes, not only the atom bomb but also the bad weather. I have endeavoured to show that the development of the human mind was bound to lead one day to the disclosure and application of the energy stored in the atomic nucleus. That this happened so quickly and so thoroughly as to lead to a critical situation is the consequence of a tragic historical accident: the discovery of the fission of uranium happened just at the moment when HITLER acquired power, and just in that country, where he acquired power. I, like many others, had then to leave Germany and I have seen with my own eyes the panic which struck the rest of the world when HITLER's initial successes made it appear possible that he might subjugate all nations of the globe. The physicists emigrated from Central Europe knew that there was no salvation if the Germans would succeed first to produce the atomic bomb. Even EINSTEIN who had been a pacifist all his life shared this fear and was persuaded by some young Hungarian physicists to warn President ROOSEVELT. Scholars emigrated from Europe contributed much to the uranium project, the most prominent of them ENRICO FERMI, perhaps the greatest experimental physicist of our time next to RUTHERFORD. The direction of the scheme remained in American hands. It seems to me that no blame can be attached to the men who constructed the atom bomb unless one accepts the teaching of extreme pacifism that power should never be used even against the greatest evil. It is quite a different matter with the application of the bombs against Japan in the last phase of the war. I personally consider this to be a barbaric act, and a foolish one. Responsible for it are not only politicians and soldiers, but a small group of scientists who advised the deciding committee appointed by President TRUMAN. One of these, FERMI, has died meanwhile. Another, from reasons of conscience, has given up all scientific activity, has become the head of a great educational institution and works against the misuse of science. Other members of this group have, as far as I know, not essentially changed their life and activities, nor presumably their opinion about the necessity of dropping the bombs on Japanese cities. If you wish to get a glimpse of the psychology of the atomic physicists read the clever and amusing book by LAURA FERMI, the widow of the physicist, *Atoms in the Family*. The title of its last chapter is 'A New Toy, the Giant Cyclotron.' This world 'toy' is significant, though perhaps overdrawn. These men are swallowed up by their problems and are triumphant if a solution is found, but ponder little about the consequences of the results. And if they do so then with the feeling: this is beyond our sphere of influence. The idea to abandon research because its effects might be dangerous seems absurd to them; for if they give up there would be plenty of others to continue, and in particular if the Americans were not on top, the Russians would be. And all, apart from a limited number, have after the war returned to peaceful occupations, to research and teaching, and they desire nothing better. Societies have been formed among them to discuss and study the social responsibility of scientists and to oppose the misuse of the discoveries.

There are of course a few physicists who have tasted power and liked it, who are ambitious and want to preserve the influential positions acquired during the war. But altogether I think that the ideal of politics without force will be less resisted by scientists than by other social groups. Even the ambitious and worldly scientists will be satisfied by directing big projects of development and advising the administrations of states in general politics. The consequences of the appearance of this type of men for the development of science itself are outside the frame of this discussion. May I be allowed to express my personal opinion that from the standpoint of fundamental research this development may turn out deplorable, perhaps disastrous. The appearance of a new EINSTEIN is hardly to be expected in such environments.

On the other hand, an admixture of scientists in politics and administration seems to me an advantage because they are less dogmatic and more open to argument than people trained in law or classics. To illustrate this let me record a recent personal experience.

There was the usual yearly gathering of Nobel Laureates, chemists and a few physicists, at Lindau, Lake Konstanz, in July, for discussing scientific problems. OTTO HAHN, WERNER HEISENBERG and myself submitted to them a declaration (called the Mainau Statement) prepared by us in collaboration with some other scholars of different countries, in which the danger of the present situation was emphasized and the abandonment of war demanded. Most of the participants agreed at once, but a few had doubts. A famous American scholar objected: 'I have just come from a visit to Israel and convinced myself that the existence of this little nation can be secured against the pressure of the Arabs only by the force of arms.' That is plausible enough. But in the end he accepted our arguments (the same as given here) and he signed the declaration with the rest of us.

Exactly the same objection is made wherever the last wars have left painful wounds, where boundaries have been shifted, populations expelled—as in Israel, Korea, Indo-China, Germany.

I myself have experienced enough to know what it means to be the victim of political persecution. I was allowed to return to my home country Germany, but my proper home land Silesia, which is now a part of Poland, is closed to me. That is a painful loss. But fate has decided. To redress the situation by force is impossible without much worse injustice and, very likely, general destruction. We have to learn resignation, we have to practice understanding, tolerance, the will to help instead of threats and force. Otherwise the end of civilized man is near.

For I believe that BERTRAND RUSSELL is right if he never tires of repeating: our choice is only between Co-existence and Non-existence. Let me end by quoting his words:

For countless ages the sun rose and set, the moon waxed and waned, the stars shone in the night, but it was only with the coming of Man that these things were understood. In the great world of astronomy and in the little world of the atom, Man has unveiled secrets which might have been thought undiscoverable. In art and literature and religion, some men have shown a sublimity of feeling, which makes the species worth preserving. Is all this to end in trivial horror because so few are able to think of Man rather than of this or that group of men? Is one race so destitute of wisdom, so incapable of impartial love, so blind even to the simplest dictates of self-preservation, that the last proof of its silly cleverness is to be the extermination of all life on our planet?—for it will be not only men who will perish, but also the animals and plants, whom no one can accuse of communism or anti-communism—I cannot believe that this is to be the end.

If we all refuse to believe this, and act accordingly, it will not be the end.

A NEW YEAR'S MESSAGE

[From *Physikalische Blätter*, Vol. 11, Jan. 1, 1955.]

Much has changed in physics during the two decades I spent abroad. It is no longer the quiet, pure science of old, but a decisive factor in the power politics of nations. I have only been a bystander of the revolution brought about by HAHN's discovery of uranium fission. It seems to me that the physicists of Germany are not as conscious of this completely changed situation as those of the Anglo-Saxon countries. There nobody can avoid the question of conscience how far he wants to collaborate in the development of forces which threaten the very existence of the civilized world. I have often asked myself how Lord RUTHERFORD, the actual founder of nuclear physics, would behave. He certainly was a patriot and helped in the defence of his country during the First World War. But he drew limits. When I came to Cambridge in 1933 FRITZ HABER was also there, ill and spiritually broken through exile from his fatherland. I tried to bring him together with RUTHERFORD; but he refused to shake hands with the originator of chemical warfare. How would RUTHERFORD behave today? He might have been able through the weight of his personality to stop the unconditional surrender of means of destruction to politicians and military. Some leading physicists of America have tried just that, but without success. There is the document in which they warned the American government not to use the atom bomb against highly populated towns and in which they predicted correctly the political and moral consequences—it is known under the name of the Franck Report after the chairman, my old friend JAMES FRANCK.

In America and England societies have been formed which aim at solving the question of the social responsibility of the scientist. As example I mention the American 'Society for Social Responsibility in Science' (S.S.R.S.), of which I am a member. This association informs its members by monthly news letters; in these we are told about discussions, talks, publications and books, and given extracts from them, also statements are published by well known men and women and finally letters from the readers are printed. In the last number there are extracts from a letter by ALBERT SCHWEITZER to the London *Daily Herald* about the hydrogen bomb and also sentences from a lecture (Alex Wood Memorial Lecture, 1954) by Professor KATHLEEN LONSDALE, the well known crystallographer, who became one of the first female members of the Royal Society. She is a Quaker and a protagonist against the misuse of scientific inventions for inhuman and political ends; she is just back from a world trip via India and Japan to Australia where she spread her ideas. She is a leader in English societies which have similar aims.

As far as I know there is no such organization yet in Germany, and that is only natural in view of the limitations which have been placed on the German scientists by the occupation statute. But the time has come when a new obligation arises from the lifting of this restraint and with it the need for clarification of these problems. It seems to me that the German Physical Society could be a forum for such discussions. It is not by any means only a matter of the most fundamental questions such as attitude towards war in general and towards the use of means of destruction, which threaten the existence of whole nations or even of all of civilized mankind. But it is also a matter of the lesser and nevertheless important problems which are concerned with the relation of the scientist to society. To select a few points:

The threatening of freedom of science by military supervision of research and censorship of publication, the spy witch-hunt as it is now rampant in the United States, the founding of numerous well-equipped state laboratories through which an increasing number of scientists fall into dependence; finally the grave question whether the successful researcher shall always remain only an expert assistant or take a responsible part in important decisions.

German physics has achieved an enormous rebuilding of her research and teaching materials in the few years since the collapse. Let her use with equal verve the perhaps only short time between now and complete freedom of action to clarify moral and social questions which have been forced on the physicist in his rôle as human being and citizen as a result of his own researches. If this is left undone the freedom of science will be as greatly threatened as the civic freedom of the individual scientist. And this problem of responsibility is as international as science herself. A uniting of the groups which discuss this in the different countries would therefore be highly desirable.

SYMBOL AND REALITY

[First published in *Universitas*, Vol. 7, No. 4, pp. 337–353 (1965).]

1. Why is Science Abstract and Mathematical?

If somebody who is no physicist, chemist or astronomer, would glance through any paper or book on these sciences, he will be struck by the amount of mathematical and other symbols and the scarcity of descriptions of natural phenomena. Even the instruments of observation are indicated only symbolically by diagrams. And yet these publications claim to deal with natural science. Where in this accumulation of formulae is living nature? How are the physical and chemical symbols connected with the experienced reality of sense perceptions?

Even the scientist himself will occasionally ponder about the reasons why he approaches nature in this abstract and formalistic way with the help of symbols. The opinion is often expressed that the symbols are just a matter of convenience, a kind of shorthand needed to handle and to master the abundance of the material. Yet I think the problem is deeper. The symbols are an essential part of the method for penetrating into the physical reality behind the phenomena. I shall try to explain this idea in the following.

2. Naïve Realism

Reality to a simple, unlearned person is what he feels and perceives. The real existence of the things surrounding him seems to be just as assured as the sensations of pain, joy or hope which he feels. He is perhaps shown an optical illusion which reveals to him that a perception may lead to doubtful or even straightforwardly wrong judgments about actual facts. But this always remains on the surface of consciousness, a curious, amusing exception.

This attitude is called naïve realism. The great majority of people preserve this state of mind throughout their life, even if they learn to distinguish between subjective experience, like pleasure, pain, expectation, disappointment, and objective experiences which have to do with things of the external world.

But there are people to whom something happens which stirs them deeply and makes them sceptical. In my case it occurred thus.

I had an elder cousin who was a University student while I was still at school. Apart from lectures on chemistry he attended also a course on philosophy which impressed him. Once he asked me suddenly: 'What do you exactly mean when you call this leaf here green or the sky there blue?' I regarded this question as rather superfluous and answered: 'I just mean green and blue because I see it like that, exactly as you see it.' But this did not satisfy him. 'How do you know that I see green exactly as you see it?' My answer 'because all people see it in the same way, of course' still did not satisfy him: 'There exist color-blind people who see the colors differently; some of them, for example, cannot distinguish red and green.' Thus he drove me in a corner and made it plain to me that there is no way to ascertain what another person perceives and that even the statement 'he perceives the same as I' has no clear meaning.

Thus it dawned upon me that fundamentally everything is subjective, everything without exception. That was a shock.

The problem was not to distinguish the subjective from the objective, but to understand how to free oneself from the subjective and to arrive at objective statements. I want to say right from the beginning that I have found no satisfactory answer to this in any philosophical treatise. But through my occupation with physics and its neighbour sciences I have arrived, near the end of my life, at a solution which appears to me to some extent acceptable.

3. Kant

At that distant time, as a young fellow, I followed my cousin and mentor who advised me to read KANT. Later on I have learned that the problem: how objective knowledge arises from the sense perceptions of the individual and what this knowledge means, is a much older one; that, e.g., PLATO's doctrine of ideas is an early formulation, followed by divers speculations of antique and mediaeval philosophers up to KANT's immediate precursors, the British empiricists LOCKE, BERKELEY, HUME. However I do not intend to speak about the history of philosophy; I only want to say a few words about KANT because he has influenced the thought of men up to our time, and because I intend to use some of his terminology.

I quote a passage from KANT's 'Critique of pure reason' (Transcendental Aesthetics): 'Objects are given us by way of sense impressions, they produce our perceptions. They are then taken up by reason which produces concepts.' Thus KANT suggests that the objects of perception are transformed by reason into concepts. He takes it as self-evident that the objects of perception are the same for all individuals and that the concepts formed by reason are moulded alike by all individuals. According to KANT all knowledge refers to the phenomena but is not determined solely by experience (a posteriori) but also by the structure of our reason (a priori). The a priori forms of our perception are space and time; the a priori forms of reason are called categories; KANT gives a catalogue of these which contains, e.g., causality.

The question whether behind the world of phenomena there is another world of objects in their own right (noumena) is, as far as I understand, left unanswered by KANT. He speaks about the 'thing in itself' but declares it as not knowable. I quote a passage from BERTRAND RUSSELL's book *Wisdom of the West* (Macdonald, London 1957); he says on page 241:

'On the Kantian theory it is impossible to experience a thing in itself, since all experience occurs with the concurrence of space, time and the categories. We may at best infer that there are such things from the postulated external source of impressions. Strictly speaking, even that is not permissible, since we have no independent way of finding out that there are such sources, and even if we had, we could still not say that they were causing our sense impressions. For if we speak of causality we are already inside the network of a priori concepts operating within the understanding.'

The vague concept of 'the thing in itself' is generally considered a weak point of KANT's teaching. One has to assume something like that in order to understand how the sense perceptions and their conceptual elaborations of the single individuals can lead to objective statements valid for all individuals. But this precondition of all objective knowledge is declared by KANT to be itself not knowable.

I shall try to show how one can escape this dilemma by using scientific methods of thinking.

But first I wish to give a short survey of the attitude to this problem taken by philosophical systems after KANT.

4. Later Philosophical Systems

I cannot dwell on the prehistory of the problem how subjective experiences were transformed into objective knowledge but only remark that it has already been vaguely discussed by PLATO in his well known simile of the cave, and more thoroughly by later philosophers, in particular by the skeptical thinker DAVID HUME. The philosophers after KANT have taken very different attitudes to it.

There are some philosophical systems which admit as real only the world of the single individual, the ipse. In my youth a German book by STIRNER, *The Only one and his Property (Der Einzige und sein Eigentum),* was widely read; as the title indicates it takes this 'solipsistic' standpoint. The fact that I remember the title of the book shows that I was impressed by it.

Much more widely accepted is the opinion, apparently shared by KANT, that it is self-evident and needs no demonstration, that the sense perceptions of different individuals are identical and that the question is only to investigate this common world of phenomena. This view is taken by the so-called 'idealistic' systems which culminated in HEGEL, and of several others, among them the 'phenomenology' of HUSSERL whose lectures I attended 60 years ago in Göttingen. He taught that one could obtain knowledge by a process of the mind called 'basic contemplation' (Wesensschau). But that did not satisfy me.

The school of logical positivism which has its roots in the work of the physicist and philosopher ERNST MACH and is today widely accepted teaches a doctrine less obscure but still more radical. Only the immediate sense impressions are regarded as real, everything else, the whole conceptual world of everyday life and of science is considered to have no other purpose than to constitute logical connections between the sense impressions. The American philosopher, MARGENAU, has introduced the term 'constructs' for all that. In the most radical interpretation this theory means a denial of the existence of an external world, or at least the negation of its knowability. In practical life a follower of this doctrine would hardly behave as if there were no external world. All these theories are relying on the same assumption that the world of sensual perception is 'the same' for all individuals. What this means is left open.

The 'materialism' of the communistic block of Eastern nations calls all these theories 'idealistic' and opposes them violently. It maintains, of course without a proof, just as an axiom, the existence of a reality independent to the subject. MARX and ENGELS seem to have regarded this like the naïve realist: matter is primary, consciousness of mind is one of its manifestations. This 'mechanical materialism' however was not easily reconciled with the results of progressing physics. For here the primitive ideas about matter were dissolved and replaced by the concept of 'field' and eventually by still more abstract ideas. Therefore LENIN invented the 'dialectical materialism' where the old term 'matter' is preserved but understood in such a general way that nothing of its meaning is preserved (just as it happened with his use of other words such as 'democracy'). The fundamental axiom is 'the existence of a real, objectively knowable external world.' Since in the East LENIN's philosophy has become a kind of official religion, a problem which has occupied and worried the minds of so many thinkers has now become an article of faith guarded by the power of the State.

Now what is the opinion of the physicists, or more general of the scientists about the problem of reality?

I should think that most of them are naïve realists who do not rack their brains about philosophical subtleties. They are content to observe a phenomenon, to measure it and describe it in their characteristic slang. As long as they have to do with instruments and experimental arrangements they use ordinary language embellished by suitable technical terms as usual in every craft.

But as soon as they begin to theorize, i.e., to interpret their observations, they use another way of communication. Already in NEWTON's mechanics, the first physical theory in the modern sense of the term, there appear concept which do not correspond to ordinary things, like force, mass, energy. With the progress of research this tendency became more and more pronounced. In MAXWELL's theory of electromagnetism the concept of the field was developed which is quite outside the world of perceptible things. Quantitative laws expressed by mathematical formulae, like MAXWELL's equations, became more and more prevalent. This happened in the theory of relativity, in atomic physics, in modern chemistry. Eventually we had in quantum mechanics a case where the mathematical formalism was developed rather complete and successful before an interpretation in words of the ordinary language was found, and even today this is not finally fixed.

What is going on here? In physics the mathematical formulae are not an end in themselves, as in pure mathematics, but symbols for some kind of reality which lies beyond the level of everyday experiences. I maintain that this fact is closely connected with the question: how is it possible to obtain from subjective experiences objective knowledge?

5. Methods of Thinking in Physics

I propose to approach this problem with the methods of thinking used by the physicist. Only a minor part of these methods is derived from philosophical systems. They have just been developed because the traditional thinking of philosophers has failed when applied to modern physics. Their strength lies in the fact that they have been successful. I mean not only that they have contributed to the understanding of natural phenomena but that they have led to the discovery of new, often overwhelmingly impressive phenomena, and to human domination over nature.

The consideration which I suggest does however not come under the title of 'empiricism' which the metaphysicists look at with contempt. The rules of thought used by the physicist are not derived from experience, but are pure ideas, inventions of great thinkers. However they are tested on an extremely large field of experience. Hence I intend not to deal with philosophy of science, but to look at philosophy from the scientific standpoint. I am sure that the metaphysicists will object to it. But I cannot help it.

To begin with I shall enumerate some of these methods of thinking and discuss their origin and their successes.

1. Decidability

I suggest the expression 'decidability' for a fundamental rule of scientific thinking (although I did not find the word in the dictionary): use a concept only if it is decidable whether it can be applied in a special case, or not.

When in electrodynamics and optics of moving bodies apparently unsurmountable difficulties were met EINSTEIN discovered that these can be reduced to the assumption, that the concept of simultaneity of events at different places has an absolute significance. This he showed is not the case, due to the fact that the velocity of light used for signalling is finite; with the help of physical means one can only establish relative simultaneity with respect to a definite coordinate (inertial) system. This idea led to the special theory of relativity and to a new doctrine of space time. KANT's ideas of space and time as a priori forms of institution were thus finally refuted.

Actually doubts about this had risen much earlier. A short time after KANT non-Euclidean geometries had been discovered (by GAUSS, BOLYAI LOBATSCHEFSKI) as logical possibilities. GAUSS made an attempt to decide experimentally whether Euclidean geometry was correct by measuring the angles of the triangle formed by three German hilltops (Brocken, Inselsberg, Hohe Hagen). He did not find a deviation of the sum of angles from the Euclidean value of 180 degrees. His successor RIEMANN took up the idea that geometry is a part of the empirical reality and developed a momentous generalization in which the idea of a curved space was introduced and worked out with mathematical rigour.

In EINSTEIN's theory of gravitation, usually called general relativity, the principle of decidability was used again. He started from the experimentally well established fact that in a gravitational field the acceleration of all bodies is equal, independent of the mass. An observer in a closed box can therefore not decide whether the acceleration of a body relative to the box is due to a gravitational field or to an acceleration of the box in the opposite direction. From this simple argument the enormous structure of the general theory of relativity was developed. The main mathematical tool was RIEMANN's geometry mentioned above, applied to the four-dimensional space, which is a combination of ordinary space and time.

I mention all this to illustrate the power and fertility of the principle of decidability. Another success of this principle is quantum mechanics. BOHR's theory of the orbital motion of electrons in the atom had, after a splendid beginning, got into difficulties. HEISENBERG observed that the theory worked with quantities which were fundamentally unobservable (electronic orbits of definite dimensions and periods) and he sketched a new theory which used only concepts whose validity was empirically decidable. The new mechanics, in the development of which I participated, did away with another of KANT's categories a priori: causality. In classical physics causality was always (doubtlessly also by KANT) interpreted as determinism. The new quantum mechanics was not deterministic but statistical (a point to which I shall return). Its success in all parts of physics is beyond dispute.

I consider it reasonable to apply the principle of decidability also to the philosophical problem of the origin of an objective world picture.

II. *Comparability. Symbols*

The point from which we started was the skeptical question: how is it possible to infer from the subjective world of experiences the existence of an objective external world? Actually this inference is innate and so natural that to doubt it seems rather absurd. But the doubt exists, and all attempts of a solution, whether of the type of KANT's 'thing in itself' or of LENIN's dogma are unsatisfactory because they violate the principle of decidability.

Now the impossibility to decide, whether the green I see is the same as the green you see, is due to the attempt to agree about one *single* sense impression. No doubt that is impossible.

But already for *two* impressions of the same sense organ, e.g., two colors, there exist decidable, communicable, objectively testable statements: they refer to the comparison of the two impressions, particularly to equality or inequality. (Instead of equal or unequal it would be better to say indistinguishable or distinguishable; but such psychological finesses do not matter in this logical consideration.) There is no doubt that two individuals can agree about such comparisons. Though I cannot describe to another person what I perceive if I call a thing green, we both can find out and agree whether two leaves which seem to me of the same hue appear to him also of the same hue. Apart from 'equality' there exist other pair relations which are communicable and objective; foremost those of the type more—less, e.g., brighter—darker, stronger—fainter, hotter—colder, harder—softer, etc. But we need not to discuss these possibilities. The existence of communicable properties of pairs suffices.

In physics this principle of objectivation is known and practised systematically. Colors, sounds, even shapes are not considered single, but in pairs. Every beginner learns the so-called zero-method, for instance in optics, where an instrument is so set that a perceptual difference of two visual fields (in brightness, hue, saturation) vanishes. The reading of a scale means the observation of a geometrical 'equality,' the coincidence of the pointer and a line of the scale. A major part of experimental physics consists in this kind of scale reading.

The fact that by comparing pairs communicable, objective statements are possible has an immense importance because it is the root of speaking and writing, and of the most powerful instrument of thinking, of mathematics. I propose to use for all these means of communication between individuals the term 'symbols.'

They are easily reproducible, visual or audible signs or signals whose accurate shape is of no importance, but for which a crude reproduction suffices. If I write (or pronounce) A and somebody else also writes (or pronounces) A, each of us perceives his own A and that of the other as equal, optically and acoustically. What matters are rough equality or some similarity—the mathematician would say: the topological aspect—not particulars such as the pitch of speaking, a flourish and ornament of writing or printing.

Symbols are the carriers of communication between individuals and thus decisive for the possibility of objective knowledge.

III. *Correspondence. Coordination*

In his "Maximen und Reflexionen" (Maxims and Reflections), Insel Bücherei, 597, GOETHE says the following: 'There is some unknown regularity in the object which corresponds to the unknown regularity in the subject.'

I quote this not only because of its relation to our discussion of subjectivity and objectivity but because of the word 'correspond.' GOETHE, with his gift of divination, has used a concept which may be called 'Urbegriff' (primary concept) of all learning, knowing, understanding. I say 'primary,' translating the German syllable 'Ur' which GOETHE himself uses in many similar cases: 'primary plant' (Urpflanze) in his doctrine of metamorphoses; 'primary phenomenon' (Urphänomen) in his theory of colors. Instead of 'correspond' one now uses often the word 'coordinate' which means making things correspond.

A child learns to speak means: it learns to coordinate words and sentences to things, persons, actions, perceptions. Writing is the coordination of visual symbols to such phenomena or to the corresponding words. Counting is the coordination of the numerals 'one, two, three . . .' learned by heart to a sequence of similar things. Modern mathematics has extended this principle to infinite sets of things, in the so-called 'theory of sets' (Mengenlehre), initiated by CANTOR. He has shown, for instance, that one cannot establish such a mutual one-to-one correspondence between the points of a (finite) line and the set of all integers (1, 2. . . without an end), which means that infinite sets of different 'numbers' exist.

In geometry points in space are connected with groups of numbers, called 'coordinates.' Thus to each geometrical fact there corresponds an analytical one, i.e., a theorem in the domain of numbers. The essential feature of mathematics is not numbers but the idea of coordination.[1] There are extended and fundamental mathe-

---

[1] It seems not to be superfluous to remark here that the current ideas about the essence of mathematics are somewhat wrong. For instance, it is repeated again and again that the whole of mathematics is a tautology, i.e., self-evident if properly considered. This opinion is expressed by the distinguished biologist and Nobel Laureate, F. B. MEDAWAR, in his book *The Uniqueness of the Individual,* p. 15 (Methuen & Co., London). The truth is that mathematics begins only with the establishment and proof of theorems for infinite sets. Thus $1+2+3+4=10$ is not a mathematical theorem, but a trivial, verifiable fact. But $1+2+3+ \ldots +n = \frac{1}{2}n \, (n+1)$ for all values $n=1, 2, \ldots$ (without end) is a mathematical theorem, simple to prove but only with the help of a principle beyond ordinary logics (the so-called principle of complete induction).

matical doctrines, like group theory, where numbers play only an insignificant part. In physics the first not purely mechanical but properly physical discovery is a perfect example of coordination, namely the discovery by PYTHAGORAS that the natural intervals in musics octave, fifth, fourth, etc. correspond to the divisions of a vibrating string according to simple ratios 2:1, 3:2, 4:3 etc. There is actually a double correspondence between perceptions of ear (musical intervals), eye or muscles (length of the string) and numbers.

The measurement of the intensity of heat (temperature) with a thermometer is the coordination of the perception of heat with a geometrical quantity (the length of a mercury column, the position of a galvanometer needle) and thus again with a number (scale value).

Chemistry coordinates the substances with combinations of symbols which are abbreviations of the names of a number of elementary substances (atoms). The historical root of this procedure is the fact that by coordinating atomic weights to the symbols of elements one could read off molecular weights from the combination of atomic symbols representing it; and by coordinating valencies to the symbols of atoms one can predict possibilities of reactions. Later this elementary method of describing chemical bonds has been absorbed by the general atomic theory.

IV. *Structures*

In every field of experiences this correspondence of sense impressions with symbols has been established. It suffices for the needs of ordinary life: the words and sentences of a language, whether spoken or written, corresponding to perceptions, emotions, etc., are learned and used without being further analyzed (naïve realism). Thus the mental image of the world is formed by the ordinary human being and refined in literature.

Science goes one step further. I do not know whether what I am going to say holds for all the sciences and the humanities. I wish to speak only about the exact sciences which I know. There mathematical symbols are used, and they have a particularity: they reveal structures.

Mathematics is just the detection and investigation of structures of thinking which lie hidden in the mathematical symbols. The simplest mathematical entity, the chain of integers 1, 2, 3 . . ., consists of symbols which are combined according to certain rules, the arithmetical axioms. The most important of these is an internal coordination: to each integer there is one following it. These rules determine a vast number of structures; e.g., the prime numbers with their remarkable properties and complicated distribution, the reciprocity theorems of quadratic residues, etc. Geometry has to do with spatial structures which appear analytically as invariants at transformations. Group theory deals with structures which appear when certain sets of operations are repeated, such as the permutations of sets of letters or symmetry operations like rotations or mirror imaging, and others.

These are structures of pure thinking. The transition to reality is made by theoretical physics which correlates symbols to observed phenomena. Where this can be done hidden structures are coordinated to phenomena; these very structures are regarded by the physicist as the objective reality lying behind the subjective phenomena.

It is impossible to describe this procedure in its enormous diversity. Only one historical point of view must be stressed: since NEWTON the structures contained in differential equations have been used and become familiar. The reason is that they permit a direct connection with experiences about ordinary things in daily life. GALILEO's mechanics started from such experiences. Then NEWTON generalized the mechanical concepts in such a way that they could be applied to celestial bodies. The first optical theories used mechanical models. Space was supposed to be filled with a substance called ether which functioned as carrier of vibrations according to the laws of mechanics. Even MAXWELL discovered and discussed his field equations at first with the help of concealed mechanism. In the early days of atomic theory mechanical models were used; in the kinetic theory of gases the atoms were considered to be small elastic balls which recoil from each other and from the walls of the container.

Very slowly and against violent opposition the opinion spread that models were not only unnecessary, but even an obstruction to progress.

The first important example was HEINRICH HERTZ' treatment of MAXWELL's theory of the electromagnetic field. HERTZ cannot be called an exclusive theoretician, for to him we owe the experimental verification of the theory through his discovery of electromagnetic waves. But he regarded the electromagnetic field as an entity in its own right which ought to be described without mechanical models.

Since then the development has irresistibly proceeded in this direction. A natural phenomenon need not be reduced to models accessible to imagination and explicable in mechanical terms, but has its own mathematical structure directly derived from experience.

This change of outlook was decisive when PLANCK discovered in 1900 in an investigation of heat radiation a new constant of nature, the quantum of action. This did not fit at all into the system of Newtonian mechanics and the physical theories built on its pattern. It is true, the models of electronic motions in atoms suggested by NIELS BOHR were a micro-imitation of planetary motion. However, not all orbits were 'allowed' but only certain 'stationary' states characterized by unmechanical 'quantum condition,' and the transitions between these states, the 'quantum jumps,' followed rules which have no analogy in mechanics. When this development culminated in the establishment of quantum mechanics there was an end to mechanical models and, by the way, also to the causal description of classical physics.

Thus physical research has won a freedom necessary to handle the ever increasing amount of observations and measurements. We try to find the mathematics appropriate to a domain of experience, then we investigate its structure and regard it as representing physical reality, whether it conforms to accustomed things or not. As examples I mention the curved space of the macro-world (cosmology) and the atoms, nuclei, elementary particles in the microworld; they have little in common with our familiar surroundings.

Yet a further freedom had to be gained before physics could claim the right to call the structures images of the reality behind the phenomena.

v. *Probability*

Philosophy has always been, and still is, inclined to make final, categorical statements. Science was strongly influenced by this tendency. The early physicists, e.g., considered the determinism of Newtonian mechanics a particular merit.

But already in the 18th century the concept of probability appears in physics. In attempts to establish a molecular theory of gases observable quantities, such as pressure, were conceived as averages of molecular collisions. In the 19th century the kinetic theory of gases was fully developed, followed by statistical mechanics applicable to all substances, gaseous, liquid and solid. The concept of probability was applied systematically and built into the system of physics.

This procedure was usually justified by the human inability to handle enormous numbers of particles with rigorous methods; but the elementary process, e.g., the collision of two atoms, was assumed to obey the laws of classical, deterministic physics.[2]

After the discovery of quantum mechanics this assumption became obsolete. The elementary processes are not obeying deterministic, but statistical laws according to the statistical interpretation of quantum mechanics.

I am convinced that ideas such as absolute certainty, absolute precision, final truth, etc. are phantoms which should be excluded from science.

From the restricted knowledge of the present state of a system one can, with the help of a theory, deduce conjectures and expectations of a future situation, expressed

---

[2] The deterministic interpretation of Newtonian mechanics is actually an unjustified idealization as BRILLOUIN and I have independently shown. It is based on the idea of absolutely precise measurements, an assumption which has obviously no physical meaning. It is not difficult to write classical mechanics in a statistical form.

in terms of probability. Each statement of probability is, from the standpoint of the theory used, either right or wrong.[3]

This loosening of the rules of thinking seems to me the greatest blessing which modern science has given us. For the belief that there is only one truth and that oneself is in possession of it, seems to me the deepest root of all that is evil in the world.

6. *Application to the Problem of the External World*

Before doing the last step in these considerations I wish to recall the point of departure, namely the shock experienced by every thoughtful person when comprehending that a single sense impression is not communicable, hence purely subjective. Anybody who has not had this experience will regard the whole discussion as sophistry. In a certain sense this is right. For naïve realism is a natural attitude which corresponds to the biological situation of the human race, just as that of the animal world. A bee recognizes flowers by their color or scent and needs no philosophy. As long as one restricts oneself to the things of everyday life the problem of objectivity is an artifact of philosophical brooding.

In science however it is different. Here one has often to do with phenomena beyond everyday experience. What you see through a high power microscope, what you perceive with the help of a telescope, spectroscope or one of the various amplifying devices of electronics is incomprehensible without a theory, it must be interpreted. In the smallest domains as in the largest, in that of atoms as well as of stars, we encounter phenomena which do not resemble the accustomed aspect of our surroundings and can be described only with the help of abstract concepts. Here the question cannot be eluded whether there is an objective world, independent of the observer, behind the phenomena.

I do not believe that this question can be answered categorically by logical thinking. But it can if we make use of the freedom to regard an extremely improbable statement as wrong.

The assumption that the coincidence of structures revealed by using different sense organs and communicable from one individual to the other is accidental, is improbable to the highest degree.

This is the normal way of scientific reasoning, and apart from science, of all research. An archaeologist, e.g., who discovers in two different countries remains of pottery of similar design will conclude that this cannot be accidental but indicates a common origin.

---

3 Here I apply the logical rule of the "excluded third" (tertium non datur). The question has been investigated, in particular in connection with quantum theory, whether a 'three-valued logic' can be established where between 'right' and 'wrong' there is a third possibility 'indeterminate.' But I cannot discuss this here.

I am not afraid of identifying such well defined structures with KANT's 'thing in itself.' The objections quoted before in the formulation of BERTRAND RUSSELL, have no validity from our point of view. They consist in the following: the existence of the 'thing in itself' is postulated because one needs an external cause to understand why different individuals experience 'the same' phenomena; but the category of causality has a meaning only within the domain of phenomena. However, the concept of causality is a residue of former ways of thinking and is replaced today by the process of coordination as described before. This procedure leads to structures which are communicable, controllable, hence objective. It is justified to call these by the old term 'thing in itself.' They are pure form, void of all sensual qualities. That is all we can wish and expect.

This result is of course contradicting to the traditional conception of KANT's 'thing in itself.' HEGEL, e.g., says in the *Encyclopedia of Philosophy*, §44: 'The thing in itself . . . means the object as far as everything referring to consciousness, feeling, emotion as well as to all notions is abstracted. It is easy to see what is left—the perfect abstractum, the complete emptiness, just something from "the other world (Jenseits) . . . .'

If the object of modern physics, in particular those of atomic physics, are identified with KANT's 'thing in itself' one can agree with HEGEL that they are a 'perfect abstractum.' But that they are perfectly empty, something from a world beyond, does not fit the facts. Remember what practical use can be made of them in the production of things like engines, aeroplanes, nuclear reactors, plastics, electronic computors and so on ad infinitum. It might happen that nuclear research leads to our being transfered to 'the other world.' Yet HEGEL did not mean this and could not foresee it.

7. Return to Images

The systems of formulae in physics do not necessarily represent conceivable things, familiar through experience. Yet they are derived from experience through abstraction and continually checked by experiment. On the other hand, the instruments used by the physicists are made of materials known in ordinary life and can be described in everyday language. The results obtained with the help of these instruments, such as exposed photographic plates, tables of figures or curves, are also of this kind. The trace of droplets in a Wilson expansion chamber suggests a particle in flight; a periodical distribution of the blackening on a photographic plate suggests interferences of waves. One cannot give up such interpretations without paralysing intuition which is the source of research and rendering communication between the scientists more difficult.

Therefore physicists are bound to describe the content of their abstract formulae as far as possible in terms of ordinary language with concepts based on intuition. The specific difficulties encountered here have been studied by the Copenhagen school under the leadership of NIELS BOHR. He has shown that it is possible to describe atomic processes with the 'classical' concepts, provided one desists from investigating all properties of a physical system simultaneously. Different, mutually exclusive but complementary experimental arrangements are needed. The experimentalist has the choice which of them to employ. Thus a subjective trend is re-introduced into physics and cannot be eliminated. Another loss of objectivity is due to the fact that the theory makes only probability predictions, which produce graded expectations. From our standpoint where subjectivity is primary and the possibility of objective knowledge problematic it is not surprising that the rigorous separation of subject and object is not possible if one tries to express the mathematical formalism with the help of images.

BOHR's principle of complementarity is another new method of thinking. Discovered in physics it is applicable to many other fields. It is another loosening of traditional methods of thought which promises important results. But this leads beyond the frame of these considerations.

I wish to mention that the latest branch of physical research, the theory of elementary particles, seems to be still entirely in the abstract. Though it leads to definite observable predictions the elementary processes themselves can hardly be grasped by intuition. The content of HEISENBERG's world-formula seems to me at present an abstract 'thing in itself' without an immediate correlation with sense impressions.

Here it may be remarked that a turning towards the abstract appears to be a general trend of our time. We observe it also in art, particularly in abstract painting and sculpture. But this parallelism is only seeming. For modern painters seem to me to avoid associations and intellectual interpretations, and concentrate on appealing to the optical sensation. The physicist on the other hand uses the perceptions of the senses as material to construct an intellectual world. The word 'abstract' refers in both cases to opposite intentions.

Yet we scientists should always remember that all experience is based on the senses. A theoretician who, immersed in his formulae, forgets the phenomena which he wants to explain is no real scientist, physicist or chemist, and if he is estranged by his books from the beauty and variety of nature I would call him a poor fool. At present we have a reasonable equilibrium between experiment and theory, between sensual and intellectual reality, and we ought to watch that it is preserved.

We also have to be careful that scientific thinking in abstract terms does not extend to other domains where it is not applicable. Human and ethical values cannot be based on scientific thinking. It is true that KANT has attempted to build up an ethical system according to the model of categories, by introducing his 'categorical imperative.' But the validity of this command is not 'decidable,' in the sense of the world as defined by us. It has to be just accepted and believed in. However attractive and satisfactory abstract thinking for the scientist, however valuable his results for the material aspect of our civilization, it is most dangerous to apply these methods beyond the range of their validity, to religion, ethics, art, literature, and all humanities.

Thus my excursion into philosophy is intended to be not only an illustration of the foundation of science, but also an exhortation to restrict the scientific methods to that domain where they reasonably belong.

WHAT IS LEFT TO HOPE FOR?

[First published in *Universitas*, Vol. 8, No. 2, pp. 105–114 (1966).]

Hope is a word one hardly finds in the literature of physics. A paper starts with the planning of an experiment or with a theory based on an expectation. But there is hardly any talk of hope.

However, when I remember my actual experiences during a long scientific career, I have one inextinguishable memory: the disappointment when a result was different from what I had expected. But disappointment could only occur when there was hope.

No science is absolutely separated from life. Even the most dispassionate scientist is, at the same time, a human being; he would like to be right, to see his intuition confirmed; he would like to make a name for himself, to be a success. Such hopes are motives for his work, just as the urge for knowledge is.

During the last decade, the belief in the possibility of a clear separation between objective knowledge and the pursuit of knowledge has been destroyed by science itself. In the operation of science and in its ethics a change has taken place that makes it impossible to maintain the old ideal of the pursuit of knowledge for its own sake which my generation believed in. We were convinced that this could never lead to any evil since the search for truth was good in itself. That was a beautiful dream from which we were awakened by world events. Even the deepest sleepers awoke when, in August, 1945, the first atom bombs fell on Japanese cities.

Since then, we have realized that through the results of our own work we are completely entangled with human life, its economy and politics, with the social struggle for power among the states, and that we therefore bear a great responsibility.

In my opinion, the atom bomb was only the last link of a development that can be traced far back and that is now leading toward a crisis, possibly toward a final, devastating catastrophe. Any hope of preventing this can only be based on an understanding of the course that led us into the present situation.

It is neither given to me nor is it my business to speak about hope in an abstract philosophical way. I can only tell about my own experiences and to what expectations they led me. I would like to show by way of certain examples how technical science applied to war has led to the gradual demolition of ethical restraints up to the present situation where there are no more restraints left. From this position there is no advance in the same direction. It is only possible to halt and then, perhaps, to turn back. That is what we may hope for.

My first knowledge of the role of modern technology in war came from history lessons at school: for instance, how the needle gun helped the Prussian army to win the war against Austria in 1866, yet how the French lost the war of 1870–71 despite their possession of the superior chassepot. This shows that, at that time, technical superiority seems to have been very important but not decisive. However, the inherent ethical danger was recognized and was met with the idea of humanization of war which found its expression in the Red Cross and in the Geneva Conventions concerning forbidden weapons, protection of the civilian population, etc.

In World War I, things developed in a different way. The war began in the old fashion with marches and battles. But soon the character of the battles changed basically. The combat zones became stationary and trench warfare developed, combined with repeated attempts to break out of it through accumulation of artillery. The soldier became more and more a mere target, an object of destruction by superhuman forces supplied by technical science. The decisive factor was the power of the industry and technical inventive capacity of the hinterland.

I myself played a tiny part in this machine as a member of a military authority in Berlin, where I worked together with other physicists on the so-called sound-ranging method. This method determined the emplacement of an enemy gun by measuring the moment of arrival of the report from firing at various observation posts. Even in this small, not very important field it was evident that everything depended on the industrial situation as a whole. The instruments for precise time measurement which we demanded from the authorities to make the method more effective were refused because the industry would not spare the time, labor, and materials for such trifles. The British, however, did not economize in this case.

The view the war offered the observer who was not confused by patriotic propaganda was this: the male youth were sacrificed in a battle that was actually decided by the hinterland's technology and supply of raw materials. Even then, this seemed most immoral and inhuman to me, and I began to understand that henceforth not heroism but technology had become decisive in war, and that in human society technology and war were incompatible.

Let me mention two experiences from World War I to illustrate this, both of them connected with the name of the great chemist, FRITZ HABER. Shortly before the war broke out, he had invented a method to fix atmospheric nitrogen (as nitric acid) and had thereby created the first artificial fertilizer, saltpeter. Now this, as is well known, is also a component of gunpowder. The German general staff had seemingly provided for everything but had not considered that saltpeter came from Chile and this import was now cut off by the blockade. Without HABER's invention, the war would have been lost for Germany after six months because of lack of gunpowder. Thus, the scientific idea and the technical capability to put it to use were in this case decisive factors of world history.

The second time HABER intervened in order to break out of the stationary trench warfare and get the front moving: he invented chemical warfare—the use of poison gas (starting with chlorine, then other much more harmful gases) in order to drive the enemy out of the trenches. This method was at first successful. But its dependence upon wind and weather and the invention of the gas mask limited its effectiveness and, further, the enemy soon knew how to use it to the same or even greater extent.

Many of my colleagues took part in this work, even men of high ethical convictions. As to HABER, so to them the defense of the fatherland was the supreme commandment. As for myself, I felt a conflict of conscience. The issue was not whether gas grenades were more inhuman than high-explosive shells but whether poison, which had been considered an instrument of cowardly murder from time immemorial, should be allowable as a war weapon, for without a limitation of the allowable soon everything might be allowed. But only years later, in fact after Hiroshima, did clear convictions began to take shape in me. Otherwise, an awareness of the scientist's social responsibility would certainly have found expression in my earlier educational work and perhaps not so many of my pupils would have been ready to collaborate on the atom bomb.

That I was not alone with my doubts even during World War I was revealed to me by an experience in 1933 when I came to Cambridge in England as a refugee. I was received with great kindness, while HABER, who had been forced to emigrate despite his services to the German cause in World War I, was not welcome. LORD RUTHERFORD, the founder of nuclear physics and one of the greatest physicists of our time, declined an invitation to my house when HABER would also be there, because he did not want to shake hands with the inventor of chemical warfare. RUTHERFORD had played a great part in the technical defense of his country and was by no means a pacifist. But he drew a line beyond which an instrument of extermination was not to be considered a weapon. I believe he would have explained that, without a moral borderline for the use of weapons, there could be no limit to devastation and that this could bring about the end of civilization.

This opinion has proved right. Chemical warfare was a decisive moral defeat of humanity. Although poison gas was not used in World War II and although Geneva Conventions have prohibited it, organizations for the study and execution of chemical warfare have been created by all military powers. A state would hardly shrink from putting one of these to use if it should prove of military advantage.

After the moral restraints concerning chemical warfare were thrown overboard, there followed the collapse of the principle which had been accepted in the nine-teenth century—that states may conduct a war only against the military forces of their enemies but not against their civilian populations.

I am not a specialist in international law and have read but little of GROTIUS and his successors. Thus I cannot present a history of this principle but only my im-pressions from events I witnessed. It is clear that the civilian population has always severely suffered from wars if they lived in the combat zone. The starvation of besieged cities and of entire countries (even after the cessation of hostilities, as after World War I) also seems to have been considered 'allowable.'

This barrier broke down during World War II as a result of the development of the air force. Germany was in the lead with air raids on open cities—Warsaw and other Polish cities, then Rotterdam, Oslo, Coventry, the systematic bombard-ment of London after Dunkirk. I was in Edinburgh at that time and often heard scornful remarks from colleagues and friends about this immoral type of warfare that England would never imitate. But this expectation proved false.

How did it happen? The leading characters in the decision on the bombing warfare were two British scientists, TIZARD and LINDEMANN. The beginning of their careers was similar: after a brilliant start in research, doubts may have risen in both men whether they would belong among the peers of science. Thus, they turned to administration and politics.

TIZARD became chairman of the committee for air defense and earned great acclaim for his role in developing the radar method for air warfare at the right time, enabling the small British air force to win the famous 'battle of Britain' and thus to thwart the plan of a German invasion.

LINDEMANN's influence too was based on his work in technical warfare but even more on his friendship with WINSTON CHURCHILL. This went back to an incident in World War I when LINDEMANN proved the accuracy of a calculation on stabiliza-tion of certain airplanes by piloting one of them himself and pulling it out of a spin. After this demonstration of ingeniousness and courage, CHURCHILL had ab-solute confidence in LINDEMANN, made him his first scientific adviser, and had a peership conferred on him under the name of LORD CHERWELL.

In 1942, LINDEMANN-CHERWELL suggested using the British bomber squadrons to destroy the workers' residential areas in big German cities. TIZARD, however, believed that an air attack against military targets would be much more effective; whether the inhumanity of the first plan made a difference to him, I do not know. CHURCHILL sided with his friend CHERWELL. Later on, it became evident that CHERWELL's estimation of the damage that would be done by the thousand-bomber raids was approximately six times too high, that these air raids were not decisive for the outcome of the war, and that TIZARD had been right.

So it happened that the German cities fell, burying hundreds of thousands of civilians under their ruins. With them fell once more a moral barrier against barbarism. Evil was met with greater evil and this again with even greater: I mean the so-called superweapons that were later used by the Germans. These were the first examples of the inhuman method of killing by remote control, without personal risk and thus without personal responsibility: that is, purely technical warfare, 'pushbutton war.'

An expert on international law might illustrate this outline of the tragic history of moral decline through many more examples: for instance, from war at sea since the introduction of submarines—remember the sinking of the Lusitania in World War I.

Under the influence of technology, the parties at war have gone so far as to deliberately exterminate the civilian population and justify this as right. Let us take a look at the numerical proportions of civilians and soldiers killed in the last three big wars that were still conducted without atomic weapons.

In World War I, the total number of killed was approximately 10 million, 95 per cent of whom were soldiers and five per cent civilians. In World War II, over 50 million were killed, comprising almost equal numbers of soldiers and civilians (52 per cent to 48 per cent). During the war in Korea, of the nine million dead, 84 per cent were civilians and only 16 per cent soldiers. Whoever still believes in war as a legitimate instrument of politics and clings to the traditional ideas of a hero's death for sake of wife and child and defense of the homeland should now realize that these are fairy tales and not nice ones at that.

Nuclear weapons have driven this development to extremes, making it obvious to everyone. One cannot blame the men who at that time (1939–1945) worked on nuclear fission, because the discovery of uranium fission came from Hitler Germany and one had to assume that the Nazis would do anything to develop it into a weapon against which there was no resistance. This had to be prevented.

But when in the United States the first bomb was ready for use, Hitler Germany had already surrendered and Japan was also at the end of its breath, having already, through diplomatic channels, even asked for peace.

Then, everything happened as in the CHERWELL-TIZARD dispute. The military leaders, especially GENERAL GROVES who had headed the nuclear energy project, thought only of the immediate military advantages and calculated how many lives would be saved if Japan were forced to surrender without an invasion. Japanese lives, of course, were not counted. GROVES, moreover, did not want to sacrifice the satisfaction of demonstrating 'his' achievement in all its horror to the world. He did not even let the scientists who had actually made the achievement have any say in the matter. Among the latter was a group of reasonable men who accurately predicted the long-term consequences of dropping this bomb on Japanese cities and, in the so-called Franck Report, warned the government accordingly. The contrary decision, however, was made by a committee appointed by PRESIDENT TRUMAN to advise him, and whose members included several outstanding physicists. These members acted according to CHERWELL's example and thus the borderline was finally crossed that leads onto the downhill road to possible self-annihilation of the human race.

I have only to indicate briefly what followed: attempts to put nuclear technology on an international basis failed. Russia made up for America's lead a lot faster than was expected. The invention of the hydrogen bomb in America was very soon imitated by the Soviet Union. Then followed the development of intercontinental ballistic missiles in competition between the two great powers, with the exploration of space program serving as a cover-up. Each of the great powers now has enough nuclear weapons to annihilate the human race many times over.

The politicians know what is at stake and they maneuvre so as to maintain the balance of terror. But the balance is unstable. The people become indifferent to the danger because of the spreading moral paralysis; the politicians become more cynical and risk advances again and again that can tip the balance—as in the Cuban crisis some time ago.

What is there left to hope for? Can one hope that the insight of mankind into the atomic danger will bring salvation?

The only thing that can save us is an old dream of the human race: world peace and world organization. These were regarded as unattainable, as utopian. It was believed that human nature is unchangeable and since there had always been war there would always be war.

Today one cannot accept this any longer. World peace in a world that has become smaller is no longer utopia, but a necessity, a condition for the survival of the human race. The opinion that this is so spreads farther and farther. The immediate result is a paralysis of politics, because a convincing method of achieving political goals without a threat of force, with war as a last resort, has not yet been discovered.

Many wise people think about this problem and I am convinced that a solution could be found if there were plenty of time at our disposal. This expectation is based on the experiences of a long life. Innumerable are the things that now exist but that in my youth were considered utopian. My field of science, atomistics and electronics, which have now led to a deep understanding of the structure of matter, were then in their very beginnings. Anyone who would have described the technical applications of this knowledge as we have them today would have been laughed at. There were no automobiles, no airplanes, no wireless transmission of communications, no radio, no cinema, no television, no assembly line, no mass production, and so on. All this has come into existence since my youth and has created economic and social changes in the lives of the people which are deeper and more fundamental than anything that has happened in the previous ten thousand years of history: semifeudal empires have become socialistic republics, a confusion of Negro tribes have become organized states with modern constitutions, space research has begun and nobody gets excited about the most daring, most expensive plans of the astronauts.

But for the one question, the most important by far, 'can political, economical, ideological disputes only be decided through force and war?' the theorem of the unchangeability of human nature is to be accepted: 'Since it has always been like this, it will always be like this.'

To me, this seems absurd even when it is preached by great politicians and philosophers. Without giving up this axiom, the human race is condemned to destruction. Our hope is based on the union of two spiritual powers: the moral awareness of the unacceptability of a war degenerated to mass murder of the defenseless and the rational knowledge of the incompatibility of technological warfare with the survival of the human race.

The only question is whether we have enough time to let these realizations become effective. For the present situation is highly unstable and becomes through its own mechanisms increasingly dangerous day by day. The failure of an individual or an apparatus, the blind passion of a leader, the ideological or national delusion of the masses can at any moment lead to a catastrophe. Perhaps we will not be spared a terrifying event before a change in thinking and action takes place.

But we must hope. There are two kinds of hope. If one hopes for good weather or for winning a pool, then hope has no influence whatsoever on what happens, and if it rains or if we draw a blank we have to resign ourselves to these facts. But in the coexistence of people, especially in politics, hope is a moving force. Only if we hope do we act in order to bring fulfillment of the hope closer. We must not tire in fighting the immorality and unreasonableness which today still govern the world.

I would like to recite here the words of a great man, not a politician or philosopher but a man of practical reasoning, the physician and Nobel Prize winner GERHARD DOMAGK, whose chemotherapeutic discoveries have preserved the health and life of countless individuals. It is as he says: 'A confession, a beseeching that is at the same time a warning and yet full of hope' and it goes as follows: 'What is really important in this world? That we individuals get along with each other, try to understand and help each other as best we can. For us physicians that is natural. Why shouldn't it also be possible for all other people? Don't tell me this be Utopia! Every discovery was considered utopian. Why should we first wait for another measuring of powers—we really did suffer enough to have become wise. But it is comfortable to cling to old customs; more comfortable to follow violent rulers, cholerics, paranoiacs, and other mentally disturbed individuals instead of thinking for oneself and looking for new ways of reconciliation instead of mutual destruction.'

These are clear words. They come from a man who fights for life, from a man who does not merely hope to save human beings from sickness but puts all his knowledge and ability to work for it.

It depends on us, on each individual citizen in every country of the world, for a stop to be put to the existing nonsense. Today, it is no longer the cholera or plague bacillus that threatens us, but the traditional, cynical reasoning of politicians, the indifference of the masses, and the physicists' and other scientists' evasion of responsibility. That which they have done, as I tried to explain, cannot be undone: knowledge cannot be extinguished, and technology has its own laws. But scientists could and should use the respect that they gain through their knowledge and ability to show the politicians the way back to reasonableness and humanity, as the Göttingen Eighteen tried once. [In 1957, eighteen distinguished German scientists, including PROFESSOR BORN, signed a declaration urging the Federal Republic of Germany to renounce the development or possession of atomic weapons of any kind. The full text of their declaration appears in the June 1957 Bulletin of the Atomic Scientists.]

All of us must fight against official lies and encroachments; against the assertions that there is protection from nuclear weapons through shelters and emergency regulations; against the suppression of those who enlighten the public about it; against narrowminded nationalism, 'gloire,' passion for great power; and we must especially fight against those ideologies which pronounce the infallibility of their doctrines and thus separate the world into irreconcilable camps.

There is still hope, but it will only come true if we stake everything on the battle against the diseases of our time.

IN MEMORY OF EINSTEIN

[First published in *Universitas*, Vol. 8, No. 1, pp. 33–44 (1965).]

I have talked about EINSTEIN many times before. That I speak once more on the same subject is due to the fact that in the leisure of my old age I have looked through the letters which Einstein wrote to me during his lifetime. There are more than fifty of them, short and long. I copied them all by hand to make doubly sure of their preservation, and this made me see my friend so vividly in the flesh that I believed to hear his voice and his wonderful laughter.

When COUNT BERNADOTTE asked me to give a general lecture in Lindau, I had the idea of trying to convey to this assembly an impression of the memories recalled by those letters. I intend to quote and discuss mainly passages concerned with questions of philosophy and physics, but occasionally I shall include also a character-istic remark about some topical question. Politics proper, though they played an important part in EINSTEIN's life, shall be left out, because they would not fit into the framework of this meeting. EINSTEIN kept all my letters and those from my wife. When the whole correspondence will be published some day, those traits of his character which I am not mentioning here, will come to light as well.

A long time before I read EINSTEIN's famous paper of 1905, I knew the formal, mathematical side of the special theory of relativity through my teacher HERMANN MINKOWSKI. Even so, EINSTEIN's paper was a revelation to me which had a stronger influence on my thinking than any other scientific experience. I made EINSTEIN's personal acquaintance in 1909 at the Congress of the German Society of Scientists and Physicians in Salzburg. I don't know whether we corresponded thereafter, because no letters from that time have been preserved.

As is well known EINSTEIN was called to Berlin in 1913 as the successor of VAN'T HOFF at the Berlin Academy, and a year later I was appointed to a chair of theoretical physics at the Berlin University to relieve PLANCK in his teaching.

We moved to Berlin in the spring of 1915, and I started a course of lectures. But I soon had to stop and join up. After a few months with the Air Force I was transferred to the Artillery Testing Commission, a military board where a group of physicists, headed by my friend RUDOLF LADENBURG, were working on the development of technical reconnoitering methods for the artillery. The board's

---

This paper was read at the 1965 meeting of Nobel Laureates at Lindau, Lake Constance, and is published here by kind permission of the author, who is known to the readers of "Universitas" by several previous contributions.

building in Spichernstraße was quite close to EINSTEIN's flat in the so-called Bavarian Quarter of Berlin. That is how I came to visit him often during the lunch break. Soon he also came to see us at home; we made music and had lively discussions, in which my wife took part. Our political views had much in common. But as I said, I am not going to talk about that.

The first written message in my collection is a postcard from his flat to mine, referring to an article of mine published in the "Physikalische Zeitschrift" in 1916, a brief account of the general theory of relativity. I would not treat the subject much differently today. It has become the fashion to regard EINSTEIN's relativistic starting-point, i.e., that the strength of a gravitational field in a box is relative to the acceleration of the box, as of secondary importance, and the field equations of the metric as the main thing. I do not like this attitude of which my Russian friend FOCK is a champion, and I still prefer EINSTEIN's original presentation as summarized in that article 50 years ago. EINSTEIN's postcard says about my article that he read it 'with the happy feeling . . . to have been completely understood and appreciated.' There follow some more extremely friendly words, and I believe this was the beginning of our friendship.

Incidentally I did not—neither then nor later—take part in working in general relativity. I thoroughly studied EINSTEIN's great papers in the Transactions of the Berlin Academy. They seemed to me so much above anything that I believed myself capable of achieving that I decided never to work in this field. But I have always upheld his views and defended them against attacks.

In the summer of 1918, EINSTEIN went on holiday with his second wife to the seaside resort of Ahrenshoop. From there I have several letters, and I shall now quote a passage from one of them (undated):

'One of the books I am reading here is KANT's Prolegomena, and I am beginning to grasp the immense suggestive power which emanated from the fellow and still does. As soon as you concede him the existence of a-priori synthetic judgments, you are caught. I must qualify the 'a-priori' and turn it into 'conventional' in order not to have to contradict him, but even so it won't fit the details. Anyway, it makes very nice reading, although he is not as good as his predecessor HUME, who also had considerably more sound instinct . . . .'

I find it refreshing to see one of the great heroes of German philosophy referred to as 'the fellow.' From such remarks I have learned the lack of respect with which you have to face philosophic thoughts if you want to achieve anything in theoretical physics. I have tried to pass this attitude on to my pupils and, as I believe, not without success.

These somewhat flippant remarks are frequent. In a letter of 1919 I apologized for the delay in my correspondence because of literary commitments. EINSTEIN wrote back: 'So you intend to keep even your literary promises—for instance to SOMMERFELD? That is going too far. If SHAKESPEARE had lived under present conditions, he would have changed his line "At lovers perjuries, they say, Jove laughs," which after all is a little hard, to "At a forgotten promise of a review." '

In the same letter there is the following observation on physics: 'The quantum theory rouses in me feelings very similar to your own. One really ought to be ashamed of the successes because they have been earned according to the Jesuit maxim "Let not thy left hand know what the other doeth." ' This is a very good characterization of the way in which at the time, before quantum mechanics, people used to juggle with the concepts of classical mechanics and quantum theory. Then, in the same letter, there follows a sermon against my political pessimism which was probably caused by newspaper reports on the peace negotiations at Versailles. EINSTEIN wrote: 'Is a hard-boiled X-brother[1] and determinist allowed to say with tears in his eyes that he has lost his belief in mankind? It is just the emotional behaviour of men in political matters today that is apt to revive the belief in determinism . . . .'

Here is the first allusion to his deterministic creed, although not in connection with physics where any doubt of strict causality would have struck him as crazy, but in relation to men's political behaviour. A detailed explanation of his belief in causality as well as profound remarks about its limits are contained in a letter to my wife. The letter was sent to Frankfurt, where I had been called in 1919 to become MAX VON LAUE's successor.

The passage reads: 'Now for philosophy. What you call "Max's materialism" is simply the causal way of looking at things. This view always answers the question "Why?", but never the question "To what end?" No utility principle and no natural selection will make us get over that. So if someone asks: "To what end shall we help one another, make life easier for each other, make beautiful music and try to produce fine ideas?", then he would have to be told: "If you don't feel it, no one can explain it to you." Without this primary feeling we are nothing, and we had better not live at all. Even if someone attempted a "justification" by trying to prove that these things help to preserve and further the existence of the human species, then the question "To what end?" would loom even larger, and providing an answer on a "scientific" basis would be an even more hopeless task. So if we want to proceed in a scientific manner at any price, we can try to reduce our aims to as few as possible and derive the others from them. But this will leave you cold.—I do not agree with your pessimistic assessment of cognition. To have a clear view of

---

1 Instead of 'to calculate' we used to say 'to X,' as the letter 'x' is employed as symbol for the unknown in every calculation.

relationships is one of the most beautiful things in life. You would have to be in a very gloomy, nihilistic mood to deny that.'

He has helped my wife, as she put it in an article on EINSTEIN in the Swiss periodical 'Die Weltwoche,' to feel no longer as though stranded in an icy moonscape living among the objective scientists. She once asked EINSTEIN: 'Well then, do you believe that it will be possible to depict simply everything in a scientific manner?' 'Yes,' said EINSTEIN, 'that is conceivable, but it would be no use. It would be a picture with inadequate means, just as if a Beethoven symphony were presented as a graph of air pressure.'

On November 9th, 1919, a short letter arrived, starting: 'Well, from now on we are going to call each other "Du" ' (the familiar address in German, in contrast to the formal 'Sie'). I need not tell you how pleased and honored I was. Sometimes it is not so easy at first for adults to get used to addressing each other as 'Du.' But in EINSTEIN's case it was simple, because he was so completely natural and frank. I don't remember if I ever relapsed into 'Sie.' In EINSTEIN's letters there are very few such relapses, and only when he was cross about something. This used to happen, mainly because of the way in which the public crowded in on him. We thought he was too gentle with officious journalists. I tried to shield him from pseudo-scientific attacks and occasionally even wrote newspaper articles in his defence. In a letter of December 9th, 1919, he wrote about such a case: 'Your article in the "Frankfurter Zeitung" pleased me very much. But now the press and other rabble are haunting you too, though to a lesser degree than me. With me it is so bad that I can hardly breathe, let alone get down to sensible work.' He then warned me against rising to the attacks of a certain person. 'Save your temperament, leave the fellow alone and let him jabber. His proof of causality a priori is truly sublime.' A report on a visit to Rostock follows in the same letter, with a funny description of the university's anniversary celebration—he never took such things very seriously. He also wrote about his visit to the philosopher SCHLICK, who later on worked in Vienna and founded the school of logical positivism, which is still flourishing today, particularly in America. EINSTEIN for a time was much impressed with the arguments of this philosophical system, but later criticized them.

On January 27th, 1920, he wrote about the quantum theory: 'I do not believe that the quanta might be understood by giving up the continuum.' (I must have made some such suggestion in a letter.) 'Analogously, one might have thought of enforcing general relativity by giving up the system of co-ordinates. On principle the continuum might well be given up. But how is one to describe the relative movement of n points in any way without the continuum? I still believe, that one has to find such overdetermined differential equations that the solutions no longer have continuum character. But how??' (There are two question marks here.) Further on in the same long letter there is an amusing remark about SPENGLER whose book *The Decline of the West* everybody was reading at the time: 'I could not escape Spengler. One sometimes likes to accept his suggestions in the evening, and smiles at them the next morning. It is evident that the whole monomania stems from schoolmaster mathematics. EUCLID—Cartesius is his contrast which he then works into everything, but—as one gladly admits—with wit. Such things are amusing, and if tomorrow someone says the reverse with the necessary wit, it will again be amusing, and the Devil only knows which is true!'

Immediately afterwards there follows something of more concern to us physicists: 'That question about causality is a great worry to me too. Can the strange features in absorption and emission of light ever be explained in terms of complete causality, or will there be a statistical remainder left? I have to confess that here I lack the courage of a conviction. But I would be most unwilling to renounce complete causality.'

In 1920, I was called to Göttingen to succeed PETER DEBYE. We had just got used to living in Frankfurt, enjoying the amenities and stimuli of the big city, and we had our doubts whether we should go to Göttingen. We asked EINSTEIN for his advice. He gave it willingly; but these considerations do not belong here. I should, however, like to mention a passage from his letter of March 3rd, 1920, which sheds some light on EINSTEIN's own life:

'After all it is not so important where you live. The best thing is to follow your heart, without thinking much about it. Also, as a man who has no roots anywhere, I don't feel qualified to give advice. My father's ashes lie in Milan. I buried my mother here a few days ago. I myself have been gadding about incessantly—a foreigner everywhere. My children are in Switzerland under conditions that entail a complicated venture for me if I want to see them. Such a man as myself considers it an ideal to be at home somewhere with his dear ones; he has no right to advise you in this matter.'

We made our minds up for Göttingen, after my efforts to get JAMES FRANCK called there at the same time had succeeded.

In a letter of January 18th, 1922, there is this passage: 'I, too, made a terrific blunder some time ago (experiment about emission of light from canal rays). But one has resign. Only death can save us from making blunders.' I am quoting this for the encouragement of the young ones who still have many blunders in front of them. The same letter contains a remark encouraging HEISENBERG and myself to go on calculating the terms of the helium atom according to the quantum rules laid down by BOHR and SOMMERFELD. We had undertaken this work in order to have a clear case of the failure of BOHR's atomic theory. Much as EINSTEIN admired this theory, he believed as little as we did that it was in any way final. EINSTEIN went on: 'The most interesting thing at present, however, is the experiment of STERN and GERLACH.' Apparently he wanted to draw my attention to it; in fact, the work had been carried out in my laboratory in Frankfurt before my very eyes.

The problem of radiation—how the wave theory could be reconciled with the concept of quanta—was occupying him all the time. In a letter of April 24th, 1924, he wrote: 'BOHR's opinion of radiation interests me very much. But I don't want to let myself be driven to a renunciation of strict causality before there has been a much stronger resistance against it than up to now. I cannot bear the thought that an electron exposed to a ray should by its own free decision choose the moment and the direction in which it wants to jump away. If so, I'd rather be a cobbler or even an employee in a gambling-house than a physicist. It is true, my attempts to give the quanta palpable shape have failed again and again, but I'm not going to give up hope for a long time yet.'

When the work on quantum mechanics by HEISENBERG, JORDAN and myself was published, he wrote to my wife on March 7th, 1926, that it had taken a hold on the imagination and thoughts of all men who were interested in theories. 'Dull resignation has given way to a tension that is unique with us thick-blooded creatures.' But my joy about this was soon damped. On December 4th, 1926, he wrote the shattering sentences: 'Quantum mechanics is most awe-inspiring. But an inner voice tells me that this is not the real thing after all. The theory gives much, but it scarcely brings us nearer to the secret of the Old Man. In any case I am convinced that He doesn't play dice.'

From the letters of the following years, dealing to an increasing degree with politics, I don't want to quote anything. When FRANCK resigned his chair in spring 1933, and I went abroad, we had EINSTEIN's full approval; he wrote: 'Thank Heavens this means no risk for either of you. But my heart bleeds when I think of the young ones,' and he writes about plans to help them, for instance by the project of a university for exiles.

EINSTEIN went to Princeton; I went first to Cambridge, then to Edinburgh. Our correspondence never stopped; it dealt with topical events as well as scientific and philosophical questions. I sent him a booklet of mine, *Experiment and Theory in Physics,* in which, stressing the primacy of experience over speculation, I attacked the wild theories of the astronomers EDDINGTON and MILNE. (Unfortunately, the booklet has never been published in German.) He wrote back on September 7th, 1944: 'I read with much interest your lecture against Hegelism (i.e., speculation) which, with us theoreticians, constitutes the quixotic element or, shall I say, the seducer? But where this evil or vice is completely absent, we are faced with the hopeless philistine. That's why I am confident that "Jewish physics" cannot be killed off.'[1] Further on in the same letter is a passage which I quoted at length in my book *Natural Philosophy of Cause and Chance* (Clarendon Press, Oxford; Dover Publications, New York) and which begins: 'In our scientific expectations we have become antipodes. You believe in the dice playing God, and I in perfect rules of law in a world of something objectively existing, which I try to catch in a wildly speculative way.'

That was the time when EINSTEIN was struggling very hard to find a 'unitary field theory,' which was to combine the fields of electricity and gravitation into one system of equations and to produce quanta and elementary particles as well. Just as it hurt me that he did not accept quantum mechanics and tried again and again to disprove it, it grieved him that his papers did not find the recognition he had hoped for. The Polish physicist LEOPOLD INFELD who for a time had worked with me in Cambridge and then joined EINSTEIN in Princeton, has recently referred to this in an autobiographical article (published in the *Bulletin of the Atomic Scientists* of February 1965). INFELD recalls that EINSTEIN told him more than once: 'Here in Princeton they regard me as an old fool.' He was looked upon as an historical relic. Yet EINSTEIN had just then embarked on a task which he carried out in collaboration with INFELD and HOFFMANN. The work was extremely difficult and important, and so bold that INFELD at first refused to believe EINSTEIN's contentions. At the time the general theory of relativity rested on two pillars. First: the movement of mass points is determined by the geodetic lines of space-time world. Secondly: the metric of this world satisfies EINSTEIN's field equations. EINSTEIN maintained that the first hypothesis was superfluous as it followed from the field equations by way of a limiting process to infinitesimally thin, mass-covered worldlines. The calculations at first were so extensive that excerpts only could be published, and the enormous manuscript was deposited in the Institute for Advanced Study in Princeton. A little later and quite independently the Russian physicist W. FOCK, whom I have already mentioned, tackled the same problem with his pupils in a slightly different way and included it in his well known book on relativity. After EINSTEIN's death his theory was presented by INFELD and PLEBANSKI in a much perfected form in their brilliant book *Motion and Relativity* (Pergamon Press, Oxford 1960). In EINSTEIN's letters of that time I only find a hint of this important investigation. A postscript to an undated letter, probably written in 1936, reads: 'INFELD is a marvellous fellow. We have done a very fine thing together: the problem of astronomic observation, considering the heavenly bodies as singularities of the field. The Institute has treated him badly. But I am going to see him through.'

---

[1] 'Jewish physics' was a Nazi term for everything connected with the theory of relativity and other abstract ideas.

Indeed, on EINSTEIN's recommendation INFELD was made professor in Toronto, but during the Cold War he was again treated badly there, and returned to his native country Poland.

The correspondence between EINSTEIN and myself now dealt with a great variety of subjects, particularly with the question of help for emigrated scientists. But the foundations of quantum mechanics were referred to again and again, for instance in the undated letter of 1936, from which I have just quoted: 'I still don't believe in the conclusiveness of the statistical interpretation of quantum theory, but I am all alone with my opinion.' I could quote many more such passages from his letters, but one shall be enough. In a letter of December 2nd, 1947, EINSTEIN admits that the theory contains a considerable amount of truth, but he goes on: 'However I cannot seriously believe in it because the theory is incompatible with the principle that physics is to represent a reality in time and space, without spookish long-distance effects.'

What the was alluding to were presumably the situations arising from the interference of probability amplitudes which, without a precise discussion on the basis BOHR's principle of complementarity strike one as somewhat paradoxical; and what is usually called 'reduction of probabilities': a state represented as a wave function in the configuration space (more generally: a vector in the Hilbert space) is turned into another one by experimental interference.

At the end of the forties, I was asked to contribute to a volume *Albert Einstein, Philosopher—Scientist,* published in the American series *The Library of Living Philosophers,* edited by P. A. SCHILPP. I also have another volume of the series, on BERTRAND RUSSELL. The books begin with a short autobiography of the scientist concerned, followed by critical essays by several authors on his fields of activity, and they end with his answer to them. I offered to write on 'Einstein's Statistical Theories.' At the end of the article I dwelled on EINSTEIN's attitude to quantum mechanics and contrasted the empiric creed of his youth (taken from an obituary on ERNST MACH) with his later inclination towards speculation. In a letter of December 3rd, 1947, he thanked me for it with these words: 'There is so much warmth in it and such a distinct proof of how odd and petrified my attitude towards the statistical quantum theory looks to you.' The same volume contains NIELS BOHR's famous report on his conversation with EINSTEIN about the philosophical problems in atomic physics, in which he refutes EINSTEIN's attempts to prove the inadequacy of a statistical interpretation of quantum mechanics by detailed discussions of examples.

But EINSTEIN did not give in. When in 1953, after reaching the age limit, I retired from my chair in Edinburgh, a Festschrift was published in my honour, containing many interesting essays. Some of them were not written in praise of me, but attacked the statistical interpretation of quantum mechanics. One of these essays was by DAVID BOHM, another by LOUIS DE BROGLIE—and a third one by EINSTEIN.

He was trying to explain his point of view by means of a simple example—a particle oscillating to and fro between two elastic walls. To me his arguments did not seem to carry conviction, particularly since I did not think the mathematical formulation of the example was correct. He was dealing with the so-called 'pure case,' in which nothing is known beyond the presence of the particle with minimum energy, whereas the corresponding classical case refers to a certain initial state of place and velocity, and in quantum mechanics has to be represented by a mixture of pure cases. This slightly more complicated problem is easy to solve; however, the transition to classical mechanics does not lead directly to a particle path with a definite initial state, but to a narrow bundle of paths. This gave me the idea to re-formulate classical mechanics in such a way that it dealt only with not sharply defined states. This presentation seems to me superior to the usual deterministic one, because the idea that there are absolutely sharp states, i.e., absolutely exact measurements, is absurd. I think that classical mechanics formulated in a statistical manner is more sensible than the usual pseudo-determinist presentation, and I hope that it will be generally adopted. Some of the 'paradoxes' of quantum mechanics appear then also in classical treatment. Instead of orbits, one has a probability distribution spreading in the phase space. Each new observation annihilates the former distribution of probability and substitutes another one; thus we have the phenomenon of the 'reduction of probabilities' that I have already mentioned and to which EINSTEIN took exception.

I sent my manuscript to EINSTEIN. The resulting correspondence is a jumble of misunderstandings, and some of his letters reveal a little irration. But I will not quote from them here. Eventually WOLFGANG PAULI, who happened to be in Princeton, intervened and tried to explain to each of us what the other had in mind. He accused me of being a 'bad listener,' which was probably justified; but otherwise he agreed with me and helped me to improve my text until he was able to approve of every word. The paper appeared in 1955 in the number of the transactions of the Danish Academy dedicated to NIELS BOHR on his 70th birthday. EINSTEIN, however, stuck to his opinion.

It was in fact a matter of a fundamental difference in our view of nature.

For years afterwards I kept turning over in my mind the philosophy which was behind my theory, and then gave a very brief summary of it in the Festschrift in honour of HEISENBERG's 60th birthday. What it boils down to is that scientific forecasts do not refer directly to 'reality,' but to our knowledge of reality. This means that the so-called 'laws of nature' allow us to draw conclusions from our limited, approximate knowledge at the moment on a future situation which, of course, can also be only approximately described. This is a way of thinking dia-metrically opposed to EINSTEIN's own, and it is not surprising that he looked upon me as a renegade. Yet I have the feeling that I have faithfully pursued the path which he showed us in his great days, while he himself stopped at a certain point. This point is the idea that the outer world as it really is, is faithfully and exactly described by science. Seen from this angle, today's theory of matter is indeed a jumble of absurdities, and EINSTEIN from his own point of view was quite right to reject it or, at most, to accept it only as provisional.

These useless and somewhat sharp discussions had no influence on our friedship and our mutual trust. I have some more letters from him which are very friendly, among them one of November 24th, 1954, congratulating me on receiving the Nobel Prize. 'I was very glad that you—even if curiously late—were awarded the Nobel Prize for your contributions to the present quantum theory. It was of course particularly your consistent statistical interpretation of the formalism which made for decidedly clearer thinking. I have not the slightest doubt about that, despite our fruitless correspondence on the subject.'

I never saw EINSTEIN again after our emigration in 1933. In his letters from Princeton he said repeatedly that he hoped to be able one day to discuss with me in person the problems that divided us. But his efforts to get me an invitation to the Institute for Advanced Study always failed—I cannot say why. Probably I was regarded there as a fossil, as he was himself, and two such relics from times past were too much for the modern masters of Princeton. EINSTEIN's last letter is type-written and only signed by him; the date is January 17th, 1955. It contained en-closed a copy of a letter to the journal *Reporter*, which was concerned with his attitude to the danger threatening the independence of scientists and the freedom of thinking during the notorious McCarthy episode. I have not dealt here with such things which are not infrequently mentioned in the letters. But from that last letter I should like to quote a few sentences: 'The paid scribes of a docile press have tried to weaken the impact of my remarks' (a warning against restricting free thought) 'by either pretending that I regretted having occupied myself with scientific aims, or by trying to create the impression that I had treated as inferior the practical professions referred to.

'What I wanted to say was only this: under present circumstances I would choose a profession in which bread-winning had nothing to do with the striving after knowledge.'

He died soon after this. My wife has a letter from his stepdaughter Margot, describing her last visit to him: 'Do you know I was lying in the same hospital as Albert? Twice I was allowed to see him and speak to him for a few hours. I was wheeled in to him in my chair. I did not recognize him at first—so changed was he by the pain and the lack of blood in his face. But his manner was the same. He was glad that I was looking a little better, joked with me and faced his own state with complete superiority; he talked with perfect calm, even with slight humor about the doctors, and was waiting for his end as if for an expected 'natural phenomenon.' As fearless as he was in life, so quietly and modestly did he face his death. He left this world without sentimentality and without regret.'

I realize what it means to have been his friend.

FROM THE
POSTSCRIPT TO 'THE
RESTLESS UNIVERSE' (1951)

CONCLUSION

We have reached the end of our journey into the depth of matter. We have sought for firm ground and found none. The deeper we penetrate, the more restless becomes the universe, and the vaguer and cloudier. It is said that ARCHIMEDES, full of pride in his machines, cried, 'Give me a place to stand, and I will move the world!' There is no fixed place in the Universe: all is rushing about and vibrating in a wild dance. But not for that reason only is ARCHIMEDES' saying pontifical. To move the world would mean contravening its laws; but these are strict and invariable.

The scientist's urge to investigate, like the faith of the devout or the inspiration of the artist, is an expression of mankind's longing for something fixed, something at rest in the universal whirl: God, Beauty, Truth.

Truth is what the scientist aims at. He finds nothing at rest, nothing enduring, in the universe. Not everything is knowable, still less predictable. But the mind of man is capable of grasping and understanding at least a part of Creation; amid the flight of phenomena stands the immutable pole of law.

> So schaff'ich am sausenden Webstuhl der Zeit
> Und wirke der Gottheit lebendiges Kleid.
> GOETHE, Faust.

> 'Tis thus at the roaring Loom of Time I ply,
> And weave for God the Garment thou seest Him by.
> (CARLYLE's translation.)

POSTSCRIPT

Since I wrote the last lines, 15 years ago, great and formidable events have happened. The dance of atoms, electrons and nuclei, which in all its fury is subject to God's eternal laws, has been entangled with another restless Universe which may well be the Devil's: the human struggle for power and domination, which eventually becomes history. My optimistic enthusiasm about the disinterested search for truth has been severely shaken. I wonder at my simplemindedness when I re-read what I said on the modern fulfilment of the alchemists' dream:

'Now however, the motive is not the lust for gold, cloaked by the mystery of magic arts, but the scientists' pure curiosity. For it is clear from the beginning that we may not expect wealth too.'

Gold means power, power to rule and to have a big share in the riches of this world. Modern alchemy is even a shortcut to this end, it provides power directly; a power to dominate and to threaten and hurt on a scale never heard of before. And this power we have actually seen displayed in ruthless acts of warfare, in the devastation of whole cities and the destruction of their population. Such acts, of course, have been achieved by other means. In the same war other cities than Hiroshima, with a considerable percentage of their population, have been destroyed a little slower by ordinary explosives. Every previous war had its technical 'progress' in destruction, back to the stone age when the first bronze weapons conquered flint axes and arrow heads. Still there is a difference. Many states, populations, civlizations have perished through superior power, but there were vast regions unaffected and room was left for new growth. Today the globe has become small, and the human race is confronted with the possibility of final self-destruction.

When the question of a new edition of this book arose I felt a considerable embarrassment. To bring it up-to-date I had to write an account of the scientific development since 1935. But although this period is as full of fascinating discoveries, ideas, theories, as any previous epoch, I could not possibly describe them in the same tone in which the book was written; namely, in the belief that a deep insight into the workshop of nature was the first step towards a rational philosophy and to worldly wisdom. It seems to me that the scientists who led the way to the atomic bomb were extremely skillful and ingenious, but not wise men. They delivered the fruits of their discoveries unconditionally into the hands of politicians and soldiers; thus they lost their moral innocence and their intellectual freedom.

On July 16, 1945, the first experimental bomb exploded near Los Alamos, New Mexico. This was certainly one of the greatest triumphs of theoretical physics if measured not by the subtlety of ideas but by the effort made in money, scientific collaboration and industrial organization. No preliminary experiment was possible, the tremendous risk was taken in the confidence that the theoretical calculations based on laboratory experiments were accurate. Therefore it is no wonder that the physicists who watched the terrific phenomenon of the first nuclear explosion felt proud and relieved from a heavy responsibility. They had done a great service to their country and to the community of allied nations.

But when, a few weeks later, two 'atomic bombs' were dropped over Japan and destroyed the crowded cities of Hiroshima and Nagasaki, they discovered that a more fundamental responsibility was on their shoulders.

The world had become pretty callous against the horrors of the war. HITLER's seed had grown. His was the idea of total war, and his bombs smashed Rotterdam and Coventry. But he found keen pupils. In the end the bombers of both sides succeeded in a systematic devastation of Central Europe. A great part of its historic and artistic treasures, the inheritance of thousands of years went up in flames. An architectural jewel like Dresden was destroyed in one of the last days of the European war, and 100,000 civilians, men, women and children, are said to have perished with it. I do not doubt that those responsible for this act can rightfully claim tactical and strategical necessity; and the world in general found sufficient justifications, ranging from blind hatred and the wish of retribution to the quasi-humane idea that to shorten the war all means are good enough. Ethical standards had fallen sharply, indeed.

Still the two atomic bombs dropped on Japan made a stir, and when details of the human tragedy became known there was something like an awakening of conscience in many parts of the world.

This is not the place to express my personal judgment of the statesmen who decided to use this brutal application of power. Cases of precedence are plentiful— there is not much difference in the responsibility for killing 20,000 in one night or 50,000 in one minute. But being a scientist I am concerned with the question of how far science and scientists share the responsibility.

The motives of those who took part in the development of nuclear explosives were certainly above reproach: many of them were just drafted to this work as their war service, others joined it, driven by the apprehension that the Germans might produce the bomb first. Yet there was no organization of scientists which could form a general opinion. Single men became little cog-wheels in the tremendous machine, which was directed by political and military authorities. The leading physicists became scientific advisers of these authorities and experienced the new sensation of power and influence. They enjoyed their work and its tremendous success, and forgot for the time being to think hard about its consequences. It is true that a group of scientists warned the U.S. Government not to use the bomb against cities, but to demonstrate its existence and power in a less murderous way, for instance on the top of Fujiyama mountain. They predicted very accurately the disastrous political consequences which an attack on a city would have. But their advice was neglected.

The principal discrepancy between public opinion in the United States and the conviction of the scientists is concerned with secrecy. The scientists are convinced that there is no secret in science. There may be technical tricks which can be kept secret for a limited period. But the laws of nature are open to anybody who is trained in using the scientific method of research.

Therefore it was futile to keep the atomic bomb project from being known to the Russian allies, and the maintenance of this secret has with necessity transformed them from old friends into enemies. They felt menaced by a tremendous new weapon; they started to develop it themselves, and they obtained it in a shorter time than was ever expected.

On the other hand this phantom of secrecy had disastrous effects on the development of nuclear physics in America. Many physicists have been subjected to suspicion and even to accusation of disloyalty. The whole of science has been hampered by the classification of discoveries into secret and open ones, and by the supervision of publication. There is no doubt that certain security measures, mainly in regard to technical questions, are unavoidable. But the subordination of fundamental research to political and military authorities is detrimental. The scientists themselves have learned by now that the period of unrestricted individualism in research has come to an end. They know that even the most abstract and remote ideas may one day become of great practical importance—like EINSTEIN's law of equivalence of mass and energy. They have begun to organize themselves and to discuss the problem of their responsibility to human society. It would be left to these organizations to find a way to harmonize the security of the nations with the freedom of research and publication without which science must stagnate.

The release of nuclear energy is an event comparable to the first fire kindled by prehistoric man—though there is no modern Prometheus but teams of clever yet less heroic fellows, useless as inspiration for epic poetry. Many believe that the new discoveries may lead either to immense progress or to equal catastrophe, to paradise or to hell. I, however, think that this earth will remain what it always was; a mixture of heaven and hell, a battlefield of angles and devils. Let us have a look around: what are the prospects of this battle, and what can we do to help the good cause?

To begin with the devil's part, there is the hydrogen bomb. We have seen that, though almost all matter is unstable in principle, we are protected against nuclear catastrophe by the low temperatures on earth, which even in our hottest furnaces are quite insufficient to initiate nuclear fusion. But the discovery of fission has destroyed this security. The temperature in an exploding uranium bomb is presumably high enough to start the fusion of hydrogen with the help of the 'carbon cycle,' which is the source of stellar energy, or a similar catalytic process. Thus an explosive of many thousand times higher efficiency than the fission bomb could be made from a material available in abundance. Of course, work has started with the usual argumentation: if we do not do it, the other fellow (meaning the Russian) will. If it succeeds there will be a new instrument of wholesale destruction, but no peaceful application of the new forces seems to be possible. No way is known to slow down fusion in order to use it as a fuel. A perfectly hellish prospect.

Fission however has many and far-reaching applications of a peaceful kind. It can be used as fuel, since the reaction velocity can be controlled. Each pile produces an enormous amount of heat which at present is wasted in most cases. Power stations using uranium or thorium as fuel are possible, as the difficulties connected with the pernicious radiation could certainly be overcome. The question is however an economic one. The raw material is rare, and if the same amount of energy which is at present made from coal would be produced by nuclear reactors, the whole uranium ore at present or in future available would be used up in less than half a century. Hence it is improbable that the new fuel will be able to compete with coal and oil. Under certain conditions, however, this may be the case, namely where the advantage of the small bulk and weight of nuclear fuel, as compared with that of coal or oil, is decisive. There is a possibility of increasing the efficiency of fission by 'breeding,' i.e., by directing the process in a pile in such a way that a great proportion of the nuclei present is transformed into fissionable isotopes. This would mean an extension of the raw material over a much longer period.

Apart from the still problematic application of nuclear reactions for power production, there are numerous others which have already led to great progress and which are more promising. There is first the generation of new isotopes in the pile. Our knowledge of the stability of nuclei and of the laws of their interaction has been immensely increased. Some of the radio-active products can be used in medicine for therapeutical purposes, replacing for instance radium in the fight against cancer. The most important application is the so-called 'tracer method' which is revolutionizing chemistry and biology. Already in the first period of radio-activity v. HEVESY had the idea to trace the fate of atoms in chemical or biological processes by adding to them a small amount of a radio-active isotope. This discloses its presence by radiation, and as the methods of detection of radiation are extremely sensitive, one can thus determine much smaller amounts of an element than with the balance. It is even possible to investigate the distribution of atoms in living tissue. The actual application of this idea was formerly restricted to the few atomic types for which naturally radio-active isotopes were known. Isotopes are now available for almost all elements of the periodic system. The work on this line, though hardly begun, has already led to important results, and will lead to still more.

But what are these important results compared with the spectre lurking in the background, the possibility of atomic warfare on a great scale?

In combination with other infernal contraptions, like rockets to deliver bombs at large distances, chemical, biological and radio-active poisons, such a war must mean a degree of human suffering and degradation which is beyond the power of imagination. No country would be immune, but those with highly developed industry would suffer most. It is very doubtful whether our technological civilization would survive such a catastrophe. One may be inclined to regard this as no great loss, but as a just punishment for its shortcomings and sins: the lack of productive genius in art and literature, the neglect of the moral teachings of religion and philosophy, the slowness to abandon outdated political conceptions, like national sovereignty. Yet we are all involved in this tragedy, and the instinct of self-preservation, the love of our children, makes us think about a way of salvation.

There are two the political colossi, U.S.A. and U.S.S.R., both pretending to aim at nothing but peace, but both rearming with all their power to defend their ideology and way of life, and between them is a weak and divided Europe, trying to steer a middle course. Both sides are greedily devouring the latest achievements of scientific technology for their armed forces. Both have some kind of theory for their way of life in which they believe with an amazing fanaticism. Yet the foundations of these theories are rather doubtful. They use the same words for different or even opposite ideas, as for instance 'democracy,' which in the West means a system of parliamentary representation freely and secretly elected, but in the East means something quite different and hard to formulate (a complicated economic and political pyramid of bureaucracy which aims at representing, and working for, 'the people'). In other ways the American theory is much vaguer than the Russian, and that seems to have a historical reason. America has grown by expansion in a practical vacuum; the pioneers of the West had to overcome terrific natural obstacles, but negligible human resistance. The Russia of today had to conquer not only natural but human difficulties: she had to break up the rotten system of the Czars and to assimilate backward Asiatic tribes; now she has set herself the task of bringing her brand of modernization to the ancient civilizations of the Far East. For this purpose it is indispensable to have a well-defined doctrine full of slogans, which appeals to the needs and instincts of the poverty-stricken masses. Thus one understands the power which MARX's philosophy has gained in the East. What can we scientists do in this conflict? We can join the spiritual, religious, philosophical forces, which reject war on ethical grounds. We can even attack the ideological foundations of the conflict itself. For science is not only the basis of technology but also the material for a sound philosophy. And the development of modern physics has enriched our thinking by a new principle of fundamental importance, the idea of complementarity. The fact that in an exact science like physics there are found mutually exclusive and complementary situations which cannot be described by the same concepts but need two kinds of expressions, can be applied to other fields of human activity and thought. Some such applications to biology and psychology were suggested by NIELS BOHR. In philosophy there is the ancient and central problem of free will. Any act of willing can be regarded on the one side as a spontaneous process in the conscious mind, on the other as a product of motives depending on past or present impressions from the outside world. If one assumes that the latter are subject to

172 PHYSICS IN MY GENERATION

deterministic laws of nature, one has a conflict between the feeling of freedom of action and the necessity of a natural process. But if one regards this as an example of complementarity the apparent contradiction turns out to be nothing but an epistemological error. This is a healthy way of thinking, which properly applied may remove many violent disputes not only in philosophy but in all ways of life: for instance in politics.

Marxian philosophy, which is a hundred years old, knows of course nothing of this new principle. However, a prominent Russian scientist has recently attempted to interpret it from the standpoint of 'dialectic materialism,' which teaches that all thinking consists of a thesis opposed by an antithesis; after some struggle, they are combined in a synthesis. In this Marxian dogma, so he claims, you have the prediction of what has happened in physics, for instance in optics: NEWTON's thesis that light consists of particles was opposed by HUYGENS' antithesis that it consists of waves, until both were united in the synthesis of quantum mechanics. That is all very well and indisputable, though a little trivial. But why not go further and apply it to the two competing ideologies: Liberalism (or Capitalism) and Communism, as thesis and anti-thesis? Then one would expect a synthesis of some kind, instead of the Marxian doctrine of the complete and permanent victory of communism. It can hardly be expected that the ideas of MARX, developed about 100 years ago, can throw much light on the development of modern science. The opposite is more likely: that the new philosophical ideas developed by science during these 100 years may help towards a deeper understanding of social and political relations. Indeed, we find two systems of thought which deal with the same structure, the state, in completely different, apparently contradictory ways. One starts from the freedom of the individual as the basic conception, the other from the collective interest of the community.

This distinction corresponds roughly to the two aspects of the problem of willing which we have just mentioned: the subjective feeling of freedom on the one hand, the causal chain of motives on the other. Thus the West idealizes political and economical liberalism, the East collective life regulated by an all-powerful state. But as it seems likely that the contradiction in the problem of free will can be solved by applying the idea of complementarity, the same will hold for the contradiction of political ideologies. Thus the intellectual gulf between West and East may be bridged, and that is the service which natural philosophy can offer in the present crisis.

The world which is so ready to use the gifts of science for mass destruction would do well to listen to this message of reconciliation and cooperation.